Oil & Gas Taxation in Nontechnical Language

Oil & Gas Taxation in Nontechnical Language

by

Frank M. Burke, Jr.
Managing General Partner
Burke, Mayborn Company, Ltd.
and
Executive Consultant
Coopers & Lybrand
Dallas, Texas

and

Mark L. Starcher
General Counsel
Tax Analysts
Washington, D.C.

PennWell Books
Tulsa, Oklahoma

Copyright © 1993 by
PennWell Publishing Company
1421 South Sheridan/P.O. Box 1260
Tulsa, Oklahoma 74101

Burke, Frank M.
    Oil & gas taxation in nontechnical language/by Frank M. Burke, Jr. and Mark L.
Starcher.
        p.   cm.
    Includes bibliographical references and index.
    ISBN 0-87814-397-1
    1. Petroleum industry and trade—Taxation—United States. 2. Gas industry—
Tazation—United States. I. Starcher, Mark L. II. Title. III. Title: Oil and gas
taxation in nontechnical language.
HD9560.8.U52B87   1993
336.2'78622338'0973—dc20   93-13053
                                                                          CIP

Printed in the United States of America

1 2 3 4 5 97 96 95 94 93

C O N T E N T S

Chapter Three 23

Types of Interests in Oil and Gas and the Tax "Property" Concept

Chapter Four 35

Acquisition and Pre-Drilling Activities

Chapter Five 49

Geological and Geophysical Costs

Chapter Six   57

Sharing Arrangements for
Exploration and Development

Chapter Seven   67

Carried Interests

Chapter Eight   71

Unitizations and Poolings

Chapter Nine   77

Gross Income

Chapter Ten   87

Intangible Drilling and Development Costs and Depreciable Equipment

Chapter Eleven   107

Operating Expenses

Chapter Twelve   113

Depletion

Chapter Thirteen 129

Tax Credits

Chapter Fourteen 135

Limitations on Deductions
and the Alternative Minimum Tax

Chapter Fifteen 145

Dispositions of Oil and Gas Properties

Chapter Sixteen   153

Types of Business Entities

Chapter Seventeen   163

International Operations by U.S. Taxpayers

Chapter Eighteen   171

Foreign Ownership of U.S. Oil and Gas Operations

Appendices   175

Index   251

ACKNOWLEDGEMENT

The authors wish to acknowledge the artistic talents of Julianne MacKinnon, a graphic artist and editor who, in her spare time, produces Tax Notes International, a leading weekly covering international tax developments. The authors hope that the wit and humor expressed in her cartoons will go a long way in leavening our tax prose.

P R E F A C E

Federal income taxation of oil and gas activities is a complex discipline that requires an understanding of both the Federal income tax system and the manner in which the industry operates. Traditional explanations of the federal income tax rules have been directed at a professional audience familiar with the underpinnings of the Federal income tax. Sadly, these tax treatises, although technically rich in detail, do not read like a Tom Clancy novel. This is understandable given the topic. A preliminary review for this book refers to the subject matter as "nauseatingly complex." Another commentator has referred to the reading of income tax treatises as akin to "inhaling dental gas," a necessary, if not soporific exercise.

Writing style aside, we felt it was important to do away with the legal jargon and explain the essence of the Federal income tax system for the person knowledgeable about oil and gas. This is especially important given the increasing knowledge gap between the day-to-day players in the oil and gas industry, and those that attempt to make sense of the byzantine rules governing the Federal income taxation of this industry.

Overview of Book

This book focuses on the Federal income tax treatment of the acquisition, exploration, development, production and disposition of oil and gas properties, and will deal with the rules relating to the transportation, refining and marketing aspects of the industry only if relevant to determining the tax consequences of production of oil and gas.

This book is written for persons having a general knowledge of the oil and gas industry. A background in accounting and general taxation is not essential, but may be helpful in understanding the concepts discussed. Executives, government officials, attorneys and accountants, academics, and students may use this book to gain a general overview of the Federal income taxation of oil and gas exploration and production activities.

Those technicians expecting to see citations to Code sections, Regulation sections, and administrative and judicial decisions will be disappointed. Citations act as shortcuts for tax professionals in expressing some thought or principle of taxation. For example, a reference to "section 61 of the Code" has a clear connotation for the tax professional, but obscures the issue (i.e., whether an item is "income" for Federal income tax purposes) for those not versed in the intricacies of tax. Since the book is directed toward those who are not oil and gas tax technicians, no citations of sections of the Internal Revenue Code (or the accompanying regulations), cases or rulings of the Internal Revenue Service (IRS) are included. This is advantageous to the casual reader in that it forces the authors to explain tax concepts in English.

Accordingly, this book should not be used as a final source in analyzing any technical question regarding oil and gas taxation. Rather, it should be used only to gain a conceptual understanding of the principles involved. If a technical question needs to be resolved, the reader should consult an attorney or accountant who specializes in oil and gas taxation. A number of well-written treatises (with citations) are also available in the field of oil and gas taxation and can be referred to for more technical discussions of the points mentioned in this book. Among the treatises dealing with oil and gas taxation are:

>Bruen, Taylor & Jensen, *Federal Income Taxation of Oil and Gas Investments* (1989).
>Dzienkowski & Peroni, *Natural Resource Taxation* (1988).
>Houghton (Ed.), *Ernst & Young's Oil and Gas Federal Income Taxation* (1991).
>Klingstedt, Brock & Mark, *Oil and Gas Taxation 1991* (1991).
>Polevoi & Smith, *Federal Taxation of Oil and Gas Transactions* (1991).
>Russell, *Income Taxation of Natural Resource* (1993).

Many of the ideas contained in these treatises are reviewed in a nontechnical manner to provide an overview to the various areas covered.

The book had its beginning in an outline prepared for the Kentucky Institute on Federal Taxation in 1978 by Frank Burke (the older author). At the time, Burke was a partner in a national accounting firm responsible for the international energy practice of the firm, and was co-author of Burke & Bowhay, *Income Taxation of Natural Resources,* which was published annually until 1985 (and is now published in updated form as Russell, *Income Taxation of Natural Resources).* The Kentucky

Institute requested Burke to present an overview of oil and gas taxation in nontechnical language instead of the usual technical presentation. This same speech was presented at the West Virginia Tax Institute in 1979 at which time Burke met Mark Starcher (the younger author), who was graduating from West Virginia University College of Law and desired to specialize in oil and gas taxation. In 1980 Starcher went to work for Burke and, since that time, the two authors have collaborated on numerous professional and academic endeavors, including several articles. The concept for this book was further refined in 1982 when the materials were revised by the authors and used for a seminar to the financial and accounting staff of the China National Offshore Oil Company in Beijing.

Since 1984, Burke has undertaken a career outside of public accounting, although he presently serves as an Executive Consultant to Coopers & Lybrand in its energy practice. Starcher has pursued a legal and publishing career in Washington, D.C. since 1982. Both authors serve on the Board of Advisors for the Research Institute of America, New York, NY.

Frank M. Burke, Jr.
Mark L. Starcher

Introduction to the Federal Oil and Gas Income Tax System

General

Oil and gas reserves are finite deposits of an exhaustible resource. An oil and gas company is, in essence, liquidating itself through production of existing reserves, unless it actively continues the search for, and exploration and development of, new oil and gas reserves. Unless replaced, when existing reserves are depleted, no significant operating assets remain.

The replacement of reserves has, over the years, become increasingly difficult and costly. Throughout the history of the oil and gas industry in the United States, meaningful financial risk has been involved in the search for oil and gas. This financial risk is increasing as attractive prospects for new domestic discoveries diminish. Since the beginning of the Federal income tax system, incentives such as depletion, the deduction for intangible drilling and development costs (IDC), and the pool of capital doctrine, have been created to temper, to some degree, the financial risk of developing oil and gas reserves. After a brief discussion of the application of the general Federal income tax rules to the oil and gas industry, these oil and gas tax incentives, their development and present status are discussed.

Gross income from oil and gas production is recognized when received by a cash basis taxpayer, or when accrued by an accrual basis taxpayer. Expenditures are generally (1) deductible in the year paid by a cash basis taxpayer, or in the year incurred by an accrual basis taxpayer, (2) capitalized as depletable tax basis of a property to be recovered over the life of the property through the cost depletion deduction or, where available, the percentage depletion deduction, or (3) capitalized as depreciable tax basis of equipment to be recovered through depreciation. In some cases, costs

Different perspectives on the same transaction.

such as certain geological and geophysical costs (G&G) are held in suspension until they can be properly classified.

If an oil and gas property is transferred to a third party, payments received by the transferor may be treated as (1) ordinary income subject to cost depletion, (2) ordinary income not subject to cost depletion, (3) a nontaxable return of capital, or (4) if treated as a sale or exchange for Federal income tax purposes, the difference between the amount received and the tax basis of the property transferred will be viewed as gain or loss from a sale or exchange. If gain results, it may be treated as capital gain or as ordinary income if depreciation, depletion, or IDC is recaptured. A loss may be a capital loss or an ordinary loss depending on the overall gain and loss situation of the taxpayer for the year in which the loss occurs.

Oil and Gas Tax Incentives

Early in the history of the Federal income tax system, the depletion concept was developed to allow capital recovery for tax purposes. In the first Federal income tax law enacted in 1913, depletion was limited to a recovery of the cost of the property, thereby encouraging the purchase of existing production in order to be able to recover the true value of oil and gas reserves as a deduction for Federal income tax purposes. Cost depletion is similar in concept to depreciation, recognizing the diminution in value of mineral reserves over time through development. Cost depletion, however, fails to compensate the oil and gas developer for exploration and development risk. The value of the deduction is tied to the cost of acquisition of existing reserves and tends to detract from new drilling and exploration. The financial risk of failure in drilling for new reserves is not adequately rewarded by simply allowing a deduction for the costs incurred.

By 1918, the need for an incentive to encourage exploration became apparent. In that year, a new provision called "discovery value" depletion was enacted. Under that concept, depletion was allowed to be based on the discovery value of an oil or gas deposit. Discovery value was the fair market value at the time of discovery. However, because of the difficulties in valuing new properties, discovery value depletion led to controversies between the Internal Revenue Service (IRS) and the industry.

By 1926, it had become clear that a different mechanism was needed. In that year, percentage depletion was enacted to replace discovery value depletion. Percentage depletion, as originally enacted, was a non-cash deduction equal to 27.5 percent of gross revenue from a property, limited to 50 percent of the net income from that property. Ironically, at the time of its enactment, percentage depletion was viewed by many as a tax reform measure that significantly curtailed the tax incentives available to oil and gas producers.

From the beginning of the income tax law, IDC were considered deductible in the year paid or incurred as ordinary and necessary business expenses expended in exploring for, and developing, oil and gas reserves. Immediate expensing of IDC was a capital formation technique that encouraged investors and oil companies to take the financial risk associated with exploration and development.

Early in the history of the income tax law, G&G were treated as ordinary and necessary business expenses deductible in the year paid or incurred. G&G include expenses incurred for surveys, mapping, and testing incurred in the search for prospective oil and gas leases. The rationale for the immediate deduction of G&G was that in order for an oil and gas company to continue as a going concern, it was "ordinary and necessary" to have a continuing search for new properties to replace depleting reserves. If G&G were not incurred on a regular, recurring basis, an oil and gas company could find itself liquidating as its existing reserves were depleted. In the early 1940s, the IRS changed its position on G&G and now requires that such costs be capitalized. This administrative interpretation creates a significant disincentive for active G&G activities, particularly for smaller oil and gas companies.

The percentage depletion and IDC deductions remained relatively unchanged from 1926 through 1969. In 1969, the percentage depletion rate was reduced from 27.5 percent to 22 percent and, in 1975, percentage depletion was repealed with certain limited exceptions. These exceptions permit percentage depletion at 22 percent on fixed contract gas and, generally, at 15 percent on a limited amount of daily production for taxpayers who qualify as "independent producers and royalty owners." Further, the minimum tax concept, first established in 1969 in the form of the add-on minimum tax, and later in the form of the Alternative Minimum Tax (AMT), has reduced the value of percentage depletion as a capital formation tool.

In 1976, certain IDC became subject to the add-on minimum tax, and later became subject to the AMT. As with percentage depletion, the effect was to substantially reduce the incentive value of IDC. Also, in recent years, certain limiting provisions, such as the "at risk" rule and the "passive loss" restrictions, have further eroded the value of the IDC deduction as a capital formation technique.

Another tax principle that developed in the oil and gas industry during the early years of the income tax system was the "pool of capital" doctrine, which allows a taxpayer to contribute cash, goods, or services to the exploration or development of an oil and gas property, and receive an interest in the property entitling the recipient to a share of the future production, without incurring taxable income or gain at the time the property interest is received. The recipient of the property interest, of course, is taxed after the property interest begins to produce income. The contributor avoids taxable income from the transaction under the pool of capital doctrine however, until the prospect becomes productive.

The pool of capital doctrine, which was judicially spawned and developed administratively by the IRS, is based on the fact that an oil and gas property represents a pool of capital to which various parties contribute during the exploration and development process. The pool of capital doctrine allows parties to share the risks of exploration and development by bringing together the capital and expertise of various parties through joint ownership of the property without Federal income tax consequences.

Even the pool of capital doctrine, which historically has been a very important capital formation tool for the oil and gas industry, has been limited in its application by the IRS and courts in recent years. While the theory may still be available for many traditional oil and gas industry transactions, its applicability is increasingly becoming limited.

Need for New Oil and Gas Tax System

While capital formation incentives are badly needed for the oil and gas industry in the 1990s, it is clear that those capital formation incentives will not be forthcoming in the tax system so long as percentage depletion and IDC remain as the primary incentives available to the oil and gas industry. These tax expenditures carry too much political baggage, and suggestions to make those provisions more attractive generally fall on deaf ears in Washington.

The industry needs to join with Congress in designing a new oil and gas income tax system that provides adequate capital formation incentives and encourages active exploration and development of oil and gas in the United States (and perhaps throughout the Western Hemisphere). For example, a system which allows all leasehold costs, G&G, equipment costs and IDC to be combined into one asset account and amortized over a period of, say, 60 months would be an acceptable substitute for percentage depletion and IDC, particularly if a tax credit were also given for a portion of the cost of exploratory wells. This system addresses the capital formation concerns of the industry in a more straightforward approach. A more detailed discussion of the need for a new tax system for U.S. oil and gas activities is found in Appendix A.

The 80-odd year history of oil and gas tax incentives has transformed the relatively straightforward concepts of IDC and percentage depletion into a series of traps for the unwary. A welcome side effect of a new system would be a more straightforward approach to taxation that, hopefully, would obviate the need for a book such as this.

Until the industry and Congress agree that a new, less complex system is essential to the health of the U.S. oil and gas industry, the Federal income tax system will continue to fall short in providing the incentives needed to assist the industry in raising the capital necessary to fund domestic oil and gas production.

A Brief Review of Oil and Gas Operations

General

Since the beginning of the Federal income tax system in 1913, the tax treatment of oil and gas exploration, development and production has been a complex, and often misunderstood, area. The tax treatment of oil and gas differs significantly from the taxation of other industries, and even other natural resources. To gain a basic understanding of oil and gas taxation (and its inherent complexity), the activities in each phase of the industry cycle: acquisition, exploration, development, production and disposition, must be reviewed

Acquisitions

Oil and gas are generally located in what geologists refer to as structural and stratigraphic traps. These traps are configurations of reservoir rocks that enable the oil and gas fluids to coalesce. In many formations, oil and water, or gas and water, are found together. Since oil floats on water, and gas is lighter than water, oil and gas deposits are found at the top of a particular trap. For an oil and gas reservoir to be commercially productive, the oil or gas must be trapped in a porous and permeable rock formation and have the necessary gravity and viscosity for commercial production.

Oil and gas may occur at any distance underground, but typically will be found thousands of feet below the surface. With current technology, geologists cannot identify with any significant degree of certainty whether commercially marketable quantities of oil or gas exist in a given location without actually drilling a well. A study

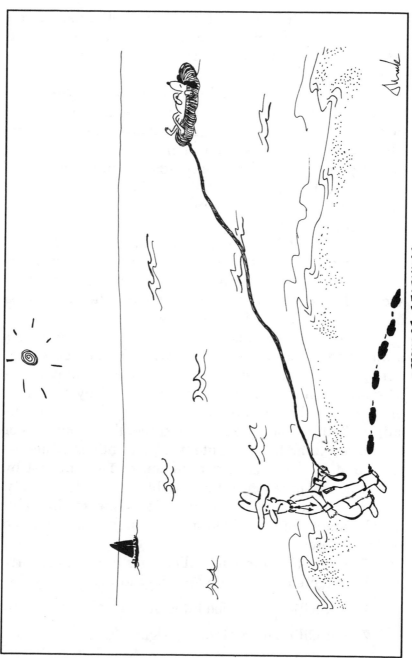

of surface conditions provides only preliminary information. A number of sophisticated methods have been developed to study onshore sub-surface conditions, including core drilling, gravity meters, seismic surveys, magnetic surveys and aerial photography from satellites. Offshore activities involve similar techniques except that controlled detonations or oscillators are used to replace the explosives customarily used in onshore exploration. Nevertheless, only actual drilling of a well penetrating the area expected to contain oil or gas deposits will determine whether oil and gas exists in commercial quantities.

Simply stated, oil and gas exploration is a search for reservoir traps. In order to secure the right to drill for, and produce, oil and gas, the rights to drill must be acquired from the owner of the mineral rights in the area in which deposits are believed to exist.

In the United States, surface rights and the minerals underlying the surface may be owned by one person or entity. This differs from the property laws of other countries where mineral rights may be severed by operation of law from the surface ownership rights. The ownership of both the surface and the minerals by one owner is referred to as the "fee interest." The mineral interest can normally be separated from the surface interest by the owner. Separation of these rights creates two assets: (1) the mineral interest which allows the exploration, development, and production of minerals, and (2) the surface interest.

The right to explore for oil and gas under a property may be obtained in several ways, including:

1. acquisition of the fee interest,

2. acquisition of exploration rights, or

3. acquisition of an oil and gas lease.

By purchasing the fee interest, an oil and gas company has outright ownership of both the surface and the mineral. However, it is frequently impossible or impractical to acquire the fee interest. Further, since the holder of the mineral interest has a right to make reasonable use of the surface to carry on exploration and development activities, there is typically no reason for an oil and gas company to acquire the surface interest.

In many cases, an oil and gas company will acquire a legal right to explore for oil and gas on a property through "shooting rights" which allow G&G work to be done on the mineral interest under a tract of land. In some jurisdictions, if the surface and mineral interests are owned by different parties, an oil and gas company must acquire shooting rights from both owners. Shooting rights are normally acquired on a per-acre basis, and provision is made for the payment of crop and other damage that may occur to the surface during the exploration activity. In most cases, an oil and gas company will acquire a "shooting option" which allows it to acquire an oil and gas lease on the property along with the shooting rights. The company may pay a per-acre dollar amount for both the right to explore (i.e., perform G&G work only) and the option to lease the property for drilling within a specified period of time. In other cases, a "shooting and selection" lease will be acquired that permits an oil and gas company to conduct G&G work and then lease certain portions of the area for drilling. This type of agreement is needed to protect the oil and gas company in the event favorable information is obtained from its G&G work.

If an area shows favorable G&G information, an oil and gas company will enter into an oil and gas lease that allows it to drill (see, for example, the sample oil and gas lease included as part of Appendix B) in the area. An oil and gas lease defines, by contract, the rights and obligations of the "lessor" (the owner of the mineral interest) and the "lessee" (the oil and gas company). The lease gives the lessee the right to explore for oil and gas during a specified period of time referred to as the "primary term" of the lease that may be for any specific time, but normally will be between three and five years. The lease also provides that it will continue for so long after the primary term as production is being obtained from the leased area. If drilling or production is not commenced by the end of the primary term, the lease will terminate, and the lessee's interest in the minerals reverts back to the lessor. Normally, the lessee may terminate the lease during the primary term by failing to pay delay rentals or by conveying the mineral interest back to the lessor.

As consideration for execution of the lease, the lessor normally receives (1) a cash "lease bonus," stated typically in a specified dollar amount per acre, (2) a promise by the lessee to pay a specified amount per acre in delay rentals, normally on an annual basis, until production begins, or until the lessee terminates the lease, and (3) a promise by the lessee to deliver to the lessor, at no cost to the lessor, a fraction of all oil and gas produced and saved from the property or, at the lessor's option, to pay the lessor the cash value of that fractional production at the wellhead. The lessor's share of production is referred to as the "landowner's royalty" or "fee royalty." Royalties and other interests in minerals having no obligation to pay drilling, development and operating expenses are referred to as "nonoperating interests." The interest acquired by the lessee under an oil and gas lease is known as the "working interest." The working interest is also referred to as an "operating interest" since it bears all obligations of drilling, developing, and operating the mineral interest subject to the lease.

If the lessee interrupts production, voluntarily or involuntarily, the oil and gas lease normally provides for payment of "shut-in royalties" to the lessor. If a successful well has been drilled, but production has not been commenced within a specified period of time, a shut-in royalty becomes payable. This type of provision applies, for example, in situations where there is no market for production from a gas well, a government permit is delayed, or no transportation is available to deliver production from the wellhead.

From time to time, oil and gas leases will also provide for "minimum" or "guaranteed" royalty payments which are a stated minimum payment required to be made by the lessee regardless of production. A guaranteed or minimum royalty may or may not be recoupable out of the lessor's share of future actual production depending on terms of the lease itself.

The landowner's royalty or fee royalty is generally stated as a fraction of production. Historically, 12.5 percent was the traditional landowner's royalty; today, however, royalties vary depending upon the relative bargaining positions of the landowner and the lessee.

While it may be thought that the relatively small fraction of production payable to the holder of the landowner's royalty is inadequate, it must be remembered that the owners of most mineral interests do not have the financial resources or petroleum expertise necessary to conduct oil and gas exploration. Further, the landowner bears no part of the exploration, development, or operating costs of the property. All costs are shouldered by the holder of the working interest. As a result, in many cases, the landowner's royalty, despite its relatively small size, may be worth as much, or perhaps more, than the working interest.

Exploration

After acquiring the oil and gas lease (or sometimes before if shooting rights are acquired prior to the actual lease), an oil and gas company will begin G&G work. As previously indicated, core drilling, seismic surveys, and similar techniques are common methods of determining the probability of the existence of oil and gas reserves. Typically, an oil and gas company will conduct detailed G&G work to determine the best location for a well.

After a location for the initial well is determined, arrangements, including all necessary governmental permitting, will be made. As part of the arrangements for drilling the initial well, an oil and gas company may attempt to share (normally using the pool of capital concept mentioned in Chapter 1) the cost of drilling with others who may benefit from the geological information to be obtained from drilling the well. Contributions to the drilling of the well may be sought from neighboring landowners or lessees. One common form of arrangement is the "bottom hole" contribution under which a neighboring landowner or lessee pays a specified amount after the well is drilled to a certain depth, regardless of whether the well is productive or dry. In other arrangements, the contributing party pays the amount due only if the well is a dry hole — a "dry hole" contribution.

Rather than incurring the cost of drilling a well directly, an oil and gas company, called the "farmor," may possibly assign the working interest in the drillsite and possible other acreage to another party, called the "farmee," under a sharing arrangement (again using the pool of capital doctrine) referred to as a "farmout." Under this type of agreement, the farmee is obligated to drill the well and normally holds the farmor's working or operating interest until all costs incurred for that well are recovered from production from the well (payout). At that point, a portion of the working interest normally reverts to the farmor. The farmee normally retains the balance of the working interest as its permanent interest in the minerals.

Farmout arrangements are quite common in the oil and gas industry and constitute one of the means of sharing drilling risks.

After the drillsite is selected and sharing arrangements, if any, are arranged, drilling begins, either by the employees of the lessee (or farmee), or by an independent drilling contractor. Independent drilling contractors normally charge (1) a flat "turnkey" cost for a well to specified conditions, (2) a per foot rate, or (3) a per day rate.

When drilling arrangements are made, site work will begin which may include road construction, drillsite preparation and excavation of mud pits. After site work is done, the drilling rig is positioned at the well site and the well is commenced, or "spudded."

Two types of drilling rigs are utilized for onshore activities. The cable tool rig, which was used in the early days of oil and gas exploration, is now obsolete. The second type, the rotary drilling rig, is now used almost exclusively. This process involves a rotating, hollow drill bit which is attached to a hollow pipe for drilling into the earth. Specially treated drilling mud is forced through the drill pipe and bit into the hole. The mud moves earth cuttings to the surface between the pipe and the walls of the hole as the drill bit proceeds. At the surface, the mud flows into the mud pits, where the drilling mud is recycled for reuse.

At the appropriate depth, surface casing is commonly set and cemented to prevent fresh water contamination and to facilitate further drilling. In deep wells, additional casing may be set to prevent cave-ins and to seal off undesirable water, gas, or oil zones.

Through the entire drilling process, samples are tested and an electronic well log records information regarding the structure of the hole.

When the projected depth of the well is reached, the well is tested. If satisfactory results are indicated, a string of casing called the "production string" is set either above the pay zone (i.e., the strata in which oil and gas are expected to flow) or through the pay zone and cemented in place. If set above the producing zone, the well is completed in the open hole below the casing. If the production string is set through the producing zone or zones, the casing will be perforated opposite the producing zone or zones. After perforations have been made in the casing (or the well is completed above the producing zone), a string of tubing is set and sealed into the well. At the ground level of the well, a "Christmas tree" is connected to the tubing. The Christmas tree is equipment designed to control and measure the flow of oil and gas to the surface where the underground pressure in the producing formation will cause the well to produce naturally. If pressure is not sufficient, a pump is installed.

A well may penetrate several producing zones or horizons. If an oil and gas company elects to produce from two or more zones or horizons at the same time, the well is referred to as a "dual completion." A well may produce oil only, gas only, or both oil and gas.

When the Christmas tree is installed, the well is considered complete. However, additional lease equipment is required for operation of the well. For example, pipe must be run from the well to separators which will separate water, oil, and gas and allow the oil to be carried to tanks. Since gas cannot be as readily stored as oil, gathering lines usually carry gas from the separator to a transmission pipeline.

To provide for efficient operation of the lease, the co-owners of the working interest normally designate a specific party (which may be one of the working interest owners or a third party) as operator of the property. Operations are generally conducted under a Joint Operating Agreement similar to the agreement found in Appendix C.

Before oil and gas can be sold, the purchaser will obtain "division orders" showing the fractional interest of each operating and nonoperating interest owner. Based on the division order, the purchaser will pay each owner its share of the sale proceeds, reduced by severance or production taxes imposed on the production. In some cases, a gas purchaser will pay the operating interest holder for all of the gas, less severance or production taxes due thereon, and the operating interest owner will pay the various other owners.

Development

After a successful oil or gas well has been discovered, an oil and gas company will typically proceed with development of the lease by drilling additional wells, which are called "development" wells. Additional sharing arrangements may be negotiated for development, or the oil and gas company may drill the development wells itself.

To maximize economic benefits from the discovery, as many wells as are permitted under pertinent state law will be drilled. It is the oil and gas company's obligation to develop the lease and produce oil or gas up to the maximum allowable under the state law, and its business interests generally will best be served by maximizing drilling.

Exploration and Development of Offshore Leases

Coastal states own land within three miles of their coastline. The area beyond this boundary, called the "Outer Continental Shelf," is owned by the United States. Federal leases normally provide for a royalty and are awarded to the oil and gas company bidding the highest lease bonus, although other leasing arrangements can be used. As contrasted with onshore exploration and development, offshore activities are more expensive and more risky. Deep water and difficult weather make offshore production activities hazardous.

Exploration and development of the offshore lease involve G&G work, much like onshore exploration, although techniques involving explosives are modified when used offshore. After an area believed to have potential is identified, mobile rigs drill exploration wells in the ocean floor, although these wells are not normally intended to be productive even if commercial quantities of oil and gas are discovered. If commercial quantities exist, the oil and gas company will manufacture and install a permanent platform for production of oil and gas. These platforms are designed to produce from several wells. Production is typically transported from the platform to shore facilities via pipeline for processing.

Production

As discussed above, revenue from oil and gas production will usually be divided and paid according to a division order. In the operation of an oil and gas lease, the owner of the working or operating interest will incur expenses, such as lease supervision and labor, utilities, repairs, ad valorem taxes (although the holders of the nonoperating interests normally pay their share of such taxes), well workover and cleaning costs, and other miscellaneous expenses of maintaining the lease and the wells thereon. If there is more than one working interest owner, one owner will normally serve as "operator" under a joint operating agreement, although, in some cases, leases may be operated by a nonowner on a contractual basis. The operator generally pays the expenses and bills for the other co-owners for their fractional share of such expenses.

The working interest owner will use all available natural forces to produce oil and gas during the "primary" production period. After natural pressure and other forces begin to fail to bring production to the surface, the working interest owner will undertake other methods, such as secondary or tertiary recovery techniques. Under secondary recovery techniques, the working interest owner attempts to increase the recoverable percentage of the oil in place by injecting into the producing formation gas or water to maintain pressure, or to cause oil and gas to move in a certain direction in the reservoir. Under tertiary recovery techniques, chemically treated substances are generally injected to attempt to recover more oil and gas. Because secondary and tertiary methods require coordination of pressure in the entire reservoir, it is beneficial to coordinate efforts with other operators in the area to maximize the efforts of the program.

Dispositions

After discovery of oil and gas on a lease, a company will typically continue to own the lease and produce oil and gas from it until it is fully depleted. At the time commercial production ceases, the lease may be abandoned.

In other cases, for economic or other reasons, an oil and gas company may decide to sell a lease during its productive life and realize the value of the lease determined from petroleum engineering reports and negotiations.

If no commercial production is obtained from a lease, the oil and gas company will typically abandon the lease. Under Federal income tax rules, the person abandoning the lease is entitled to a deduction for the basis in the mineral property at the time the property is determined to be worthless. While the most common method of abandoning a lease is ceasing to pay delay rentals, or conveying the lease back to the lessor, a taxpayer may retain title to the mineral interest and claim the Federal income tax worthless property deduction if he can provide a well-documented determination that the lease has no value.

Types of Interests in Oil and Gas and the Tax "Property" Concept

General

In a legal context, all rights to oil and gas (and solid minerals) in the gound are referred to in the aggregate as "minerals." Even if the owner of the minerals does not own the surface of the land under which the minerals are located, the mineral owner has the right to make reasonable use of the surface to develop the property for production of oil and gas.

The owner of the minerals has the right to grant and receive compensation for an oil and gas lease. This includes the right to sell interests in the minerals to more than one party. For example, the holder of 1,000 acres of mineral rights may assign all or a specific portion (such as the southwest 250 acres) to a third party. Also, the holder of the mineral rights might assign a 50 percent undivided interest in the entire 1,000 acres to a third party. It is also possible for the mineral owner to make a horizontal assignment of all, or a portion, of the oil and gas above or below a designated depth.

Mineral rights may be assigned in perpetuity, or for a limited period of time, in which case, the limited term assignment conveys "term" minerals. If the assignor of mineral rights retains a right to the bonus or delay rentals from future leasing transactions, the lessee receives a "nonparticipating" mineral interest.

As described in Chapter 2, if the owner of the mineral interest enters into an oil and gas lease, the leasing transaction results in a creation of a landowner's royalty, which is a nonoperating interest, and the working interest, which is an operating interest. The working, or operating interest, represents an interest in minerals in place burdened by the cost of exploration, development, and production. The royalty interest, or nonoperating

interest, created in the oil and gas leasing transaction is a right to oil and gas in place entitling its owner to a specified fraction, in kind or in value, of oil and gas produced from the property free and clear of the costs of exploration, development, and production.

Applicable Property Law

For Federal income tax purposes, "property" is a defined term that may not coincide with the legal status or characteristics of a mineral interest under state property laws. This definition applies to all oil and gas interests, regardless of state law, to create a uniform application of the Federal income tax law in all cases.

Creation of Other Interests in Minerals

The lessee, as owner of the working interest, may retain the entire working interest, or divide it into various other interests to finance exploration and development or spread the risk of exploration and development. For example, the owner of the working interest may transfer the working interest to a third party and retain a nonoperating interest in the conveyance through a farmout or sharing arrangement (defined in Chapter 6). On the other hand, the lessee may retain the basic working interest and create a nonoperating interest in the hands of a third party. Quite often more than one interest is created out of a single working interest.

Overriding Royalty

An overriding royalty is similar to a landowner's royalty in that each interest has a right to minerals in place which entitles the holder of that interest to a specified percentage of production, either in kind or in value with no offset for the cost of exploration, development, and production. An overriding royalty differs from a landowner's royalty in that it is created out of the working or operating interest, and its life is limited to the life of the interest from which it is created. A working interest owner may create an overriding royalty in the hands of a third party by conveyance, or may retain an overriding royalty upon the assignment of the working interest to a third party.

Net Profits Interest

A net profits interest represents an interest in minerals entitled to a share of gross production measured by the net profits (as defined in the agreement among the parties) from operation of the property. Occasionally, the landowner's royalty may take the form of a net profits interest, defining the royalty as a share of net profits, rather than a share of gross production. Ordinarily, however, a net profits interest is created by conveyance by the holder of the working interest and has a life coexistent with the working interest.

The holder of the net profits interest has no legal obligation for payment of exploration, development, or production costs if the share of production attributable to the net profits interest is not sufficient to pay those costs. The amount of the net profits interest is computed by reducing gross income accruing to the net profits interest by defined exploration, development, and production costs, and other adjustments specified by the agreement creating the net profits interest. The net profits interest, however, bears expenses only to the extent of its share of gross income, and no payment is required of the holder if expenses exceed the allocable gross income. If no net profits exist, no payment is made to the holder of the net profits interest.

Since a net profits interest is normally a percentage of the net profit realized by the holder of the working interest, it is essential that the document creating the net profits interest clearly specify the methodology to be used in calculating the net profits interest. For example, the document should specify that until any accumulated net losses attributable to the net profits interest are offset by future net profits attributable to it, the net profits interest owner receives no payments. Failure to specifically cover this point could lead to confusion since it is also possible for accumulated losses to be ignored in the calculation of a net profits interest. To further reduce the possibility of controversy, a detailed accounting procedure should be set forth in the agreement.

One item that must be carefully dealt with in the agreement is the treatment of equipment costs and depreciation. While equipment costs and depreciation are normally taken into account in calculating the net profits interest owner's share of net profits, the holder of the net profits interest does not have a depreciable interest in lease and well equipment and cannot claim depreciation on those items for Federal income tax purposes.

It is quite common for a net profits interest to be in

the form of a contract, rather than a recordable interest in real property. The holder of a net profits interest will nevertheless be considered the owner of a portion of gross production measured by net profits for Federal income tax purposes. If the net profits interest is a contract, rather than a recordable property interest, the holder should be sure that the contractual interest is protected in case of a change in ownership of the working interest to which the contractual interest relates.

Production Payment

For many years, production payments were used as a financing technique in the oil and gas industry. Since 1969, the use of production payments has diminished significantly because of changes in the tax laws. Nevertheless, production payments continue to be used in certain financing arrangements and have Federal income tax implications.

A production payment represents a right to minerals in place under which the owner of the production payment is entitled to a specified percentage of production for a limited period of time or until a defined amount of money has been received or until a defined number of units of production have been received. To qualify as a production payment, the interest must have a life shorter than the economic life of the interest from which it is created. The production payment owner bears no part of exploration, development, or production costs and has no rights except to the proceeds from production when and if produced.

The "Economic Interest" Concept

For Federal income tax purposes, it is essential that the holder of an interest in minerals possess an "economic interest" since ownership of the economic interest determines who is taxed on production income and who is entitled to the deduction for depletion on that production.

While the term "economic interest" is somewhat difficult to define, its essence lies in the ownership of minerals in place. For Federal income tax purposes, an economic interest is owned where a taxpayer has acquired, by investment, any interest in oil and gas in place and secures, by any form of legal relationship, income derived from the extraction of oil and gas to which the taxpayer must look for the return of its capital. If the holder of an interest in minerals can look to a source other than production for return of capital, such as from a guarantee by a third party or from sources unrelated to the production of oil and gas, the holder does not own, in all likelihood, an economic interest in minerals in place for Federal income tax purposes.

The holders of a landowner's royalty, working interest, an overriding royalty, or a net profits interest all typically possess an economic interest. However, the holder of a production payment possesses an economic interest only if:

1. the proceeds from the assigned production payment are pledged to the exploration or development of the mineral property burdened by the production payment, or

2. the production payment is retained by the lessor in a leasing transaction.

The Tax "Property" Concept

The term "property" has a specific meaning for Federal income tax purposes. The technical tax meaning of property is often different from the term as generally understood in the industry. Nevertheless, much of the Federal income tax law governing oil and gas operations focuses on the tax concept of "property." Simply stated, a tax "property" is the accounting unit used in applying many of the Federal income tax rules applicable to oil and gas operations. For the remainder of this book, the term "property" will mean tax "property" as defined for Federal income tax purposes.

As indicated above, the oil and gas leasing transaction creates a working, or operating, interest and a royalty, or nonoperating, interest. As a lease is explored and developed, an oil and gas company may transfer an interest in the lease to others, or transfer the lease and retain one or more interests in it. Further, as oil and gas development occurs, several leases may be involved in the overall activity. The various transactions involved in oil and gas exploration and development frequently result in the creation of two or more tax "properties." The type of property interests created from these exploration and development activities can dramatically affect the Federal income tax consequences associated with the production of the oil and gas.

The tax property definition is referred to for purposes of determining a number of tax consequences, including the following:

1. Depletion (and certain adjustments thereto) is calculated on a property-by-property basis.

2. IDC (and certain elections relating thereto) are calculated on a property-by-property basis.

3. G&G are accounted for on a property-by-property basis.

4. The property concept is important in determining the tax consequences of certain sharing arrangements.

5. The timing and amount of the worthless property deduction are affected by the property concept.

6. The amount and character of gain from disposition of a property are calculated on a property-by-property basis.

7. The Federal income tax consequences of pooling and unitization agreements are affected by the property concept.

8. The application of the at-risk rule and the calculation of the AMT utilize the property concept.

For Federal income tax purposes, a tax property is defined as each separate interest owned by a taxpayer in each mineral deposit in each separate tract or parcel of land. As a general rule, each mineral interest (working interest, royalty, overriding royalty, net profits interest, or production payment constituting an economic interest) is treated as a separate tax property, unless the same type of interest is acquired simultaneously from the same assignor in tracts or parcels of land which are geographically contiguous.

The term "separate interest" normally refers to the nature of the mineral interest held. For example, ownership in fee simple or ownership of a working interest, royalty, overriding royalty, or net profits interest will normally be treated as a separate interest. If two or more operating interests are owned within a tract or parcel of land, those interests are combined and treated as one property, unless the taxpayer elects to treat them as separate properties. An election may also be made to combine certain operating interests in a tract or parcel of land and treat others as separate, but only one such combination may be made within a tract or parcel of land. Nonoperating interests may not be included in the combination.

Two or more nonoperating interests in a single tract or parcel of land, or in adjacent tracts, may be combined with the consent of the IRS. For Federal income tax purposes, tracts are considered contiguous when they have at least some common border. If two or more tracts touch only at one corner, they are viewed as adjacent, not contiguous.

Different types of interests in a tract or parcel of land acquired at different times are treated as separate properties. A question still exists, however, whether different types of interests in the same tract or parcel of land acquired at the same time from the same owner constitute one or two properties. Despite some judicial authority to the contrary, the IRS takes the position that the simultaneous acquisition or retention of two different kinds of interests in a single tract or parcel of land results in only one property.

If two noncontiguous tracts are acquired from the same owner at the same time, these tracts will be treated as two separate properties even though the tracts contain parts of the same mineral deposits. Two separate properties will also be deemed to exist if two contiguous tracts are acquired from different owners at the same time, or from the same owner at different times. However, if two contiguous tracts are acquired from the same owner (whether in the same or separate leases) at the same time, the two tracts are one property. Offshore leases, which must be bid separately under Federal law, are separate properties even though the leases may be contiguous, and bid and acquired on the same date.

In determining the existence of a property, the number of different mineral deposits underlying the tract or tracts of land must be determined. The existence of several mineral deposits under tracts which are contiguous and acquired from the same person at the same time can result in each mineral deposit being a separate property if the owner so elects.

While the election to treat separate deposits as separate properties applies only after a horizon has been clearly identified, the actual existence of a mineral deposit may not be necessary to create a separate property. If a taxpayer can demonstrate that a horizon is anticipated and identified by G&G work, that evidence should allow an election to be made to treat the identified horizons as separate properties. This election has, however, been the subject of litigation, and no clear guidelines on the requirements for making the election for separate deposits exist at this time.

CHAPTER 4

Acquisition and Pre-Drilling Activities

General

As suggested in Chapter 2, the ownership of minerals may be achieved through a purchase, or through an oil and gas lease. Classification of the acquisition transaction is critical since the parties may realize significantly different Federal income tax consequences from the two different types of transaction.

Purchase of Mineral Rights

A purchase of minerals is normally treated as a sale or exchange by the selling party. The owner of any type of property interest may assign or sell all of his interest, or a fractional interest identical (except as to quantity or percentage) with the fractional interest retained. In addition, the owner of an undeveloped working interest will be considered to have made a sale or exchange when an assignment of any type of continuing nonoperating interest in the property is made and the working interest is retained. Further, when the owner of a continuing interest conveys that interest and retains a non-continuing interest (an interest which terminates before the end of the productive life of the property, such as a production payment), a sale or exchange has occurred.

A sale or exchange may be illustrated by assuming that the owner of a working interest conveys the entire working interest for cash.

The seller has made a sale for Federal income tax purposes because the purchaser has obtained all of the seller's rights in the working interest. If the seller conveys an undivided 40 percent interest, rather than the entire interest, the result will be the same because the purchaser's interest will be identical to that of the seller, except as to percentage interest. In addition, if the seller sells a specific tract of acreage within the working interest, that conveyance will also be deemed a sale or exchange. In each case, the seller of the undeveloped working interest will have capital gain income (unless the seller is a dealer in oil and gas properties holding the property for resale to customers in the normal course of business) equal to the excess of the sale proceeds over the seller's adjusted basis in the interest conveyed. If the working interest sold has been the subject of exploration and development activity prior to sale, a part of the gain may be ordinary income due to recapture of IDC and depletion as discussed in Chapter 10. The gain allocable to equipment may be treated as ordinary income because of depreciation recapture as discussed in Chapter 10.

The purchaser must capitalize the amount paid for the working interest as its tax basis in the property. That basis is subject to cost depletion (as described in Chapter 12) if the property proves productive. Also, some purchasers will qualify for percentage depletion (described in Chapter 12) on the property which may provide a larger deduction than cost depletion (percentage depletion, as contrasted with cost depletion, is allowed without regard to the taxpayer's basis in the property).

The owner of the working interest may divide that interest into various other interests and sell those interests to others. The interest created from the working interest must be a continuing interest in the property. For example, the working interest owner might create and sell a ten percent overriding royalty out of the working interest. Again, the taxpayer will be deemed to realize gain to the extent the sale proceeds exceed the tax basis in the interest sold. The purchaser will be required to capitalize the purchase price as its tax basis in the property acquired. If the interest created out of the working interest is not a continuing interest, such as a production payment, a sale has not occurred, but rather the cash received will usually be treated as a mortgage loan on the property.

Another example of a sale or exchange transaction is where holder of a working interest conveys that interest, retaining a production payment payable out of production from the property conveyed. The transaction is treated as a sale or exchange with the seller being deemed to have sold his entire interest, retaining a purchase money mortgage. Again, the purchaser will capitalize the consideration paid, including the production payments, as tax basis in the property acquired.

Oil and Gas Lease

A conveyance is treated as a lease (as opposed to a sale or exchange) in any case where the owner of the minerals assigns all or a part of the minerals either for no immediate consideration, or for cash or its equivalent, and retains a continuing nonoperating interest in the property. Normally, the transaction whereby the original mineral owner conveys the working interest to an oil and gas company and retains a landowner's royalty is referred to as a "leasing" transaction. A subsequent transaction where the owner of the working interest assigns the working interest to another party and retains an overriding royalty is referred to as a "subleasing" transaction.

A leasing or subleasing transaction always involves

the transfer of the working interest, with the retention of a continuing nonoperating interest, such as an overriding royalty or a net profits interest.

For Federal income tax purposes, the lessee in a leasing or subleasing transaction normally pays the lessor consideration in the form of a lease bonus. Since a bonus is in the nature of an advance royalty, the bonus payment must be capitalized and recovered through depletion over the life of the property. If the property does not become productive and the lease is abandoned, the lessee may deduct its investment as an abandonment loss. For example, if the owner of minerals grants a lease to an oil and gas company, retains a landowner's royalty and receives a $100,000 cash lease bonus from the oil and gas company, the oil and gas company will be required to capitalize the $100,000 bonus payment as leasehold cost. If the property becomes productive, those costs will be recovered through the depletion allowance as explained in Chapter 12. If the property does not become productive, the oil company will claim a $100,000 loss at the time the property becomes worthless. Since a lease bonus can only be recovered through depletion, or at the time the property is abandoned as worthless, oil and gas companies prefer to limit the amount of bonus payments.

The lease bonus is considered to be an advance royalty to the lessor in an oil and gas leasing transaction. This payment is treated as ordinary income in the year received subject to cost depletion as described in Chapter 12. The tax basis which the assignor had in the property prior to the leasing transaction becomes the tax basis in the retained nonoperating interest. The assignor is treated as having continued his interest in the property through retention of the royalty which, presumably, has a value equal to the value of the interest owned before the lease. Any cost depletion claimed on the lease bonus will reduce the tax basis in the interest retained. For example, if the owner of the minerals acquired the minerals for $50,000 and subsequently granted an oil and gas lease, retained a royalty and received a lease bonus of $100,000, the bonus would be ordinary income in the year of receipt, subject to cost depletion. The $50,000 tax basis in the minerals would be carried over as the basis of the retained nonoperating interest. The basis in the nonoperating interest would be reduced by any cost depletion claimed on the lease bonus. Obviously, a lessor desires to obtain as large a lease bonus as possible and, hence, the objectives of the mineral owner are different from those of the oil and gas company. Given the differing motivations of the taxpayers, it is wise for the parties to a transaction to agree how it should be viewed for Federal income tax purposes.

Subleasing Developed Property

A different situation exists if the working interest being conveyed is partially or fully developed before the transaction. If the property subject to the subleasing transaction includes depreciable assets, the consideration received in a subleasing transaction must be allocated between lease bonus and consideration for the depreciable assets. The IRS contends that the cash consideration must first be applied to the depreciable basis of the equipment and that any excess cash received is lease bonus unless evidence indicates that the fair market value of the equipment is more than the depreciable basis, in which case an additional part of the consideration will be allocated to the equipment, and gain may be recognized on the sale of the equipment. Further, according to the IRS, any tax basis in the equipment in excess of the cash consideration received must be transferred to the depletable tax basis of the retained nonoperating interest and recovered through depletion. Judicial authority, however, indicates that if the basis of equipment exceeds cash consideration, a loss should be recognized on the sale of equipment. The reasoning for this position is that depreciable basis cannot be converted into depletable basis. If the value of the equipment is in excess of its tax basis, and gain is recognized on the sale of the equipment, all or part of the gain may be ordinary income (instead of capital gain) because of the recapture of depreciation (see Chapter 10).

Sharing Arrangements

It is quite common to have transactions, referred to as "sharing arrangements," to spread the risk of exploration and development under the pool of capital doctrine described in Chapter 1. In a typical sharing arrangement, the assignor conveys all or some part of its interest in the minerals for a contribution to the exploration and development of the property. In these transactions, the consideration is the contractual assumption by the assignee of all or some part of the burden of exploration and development of the property.

Because of the pool of capital doctrine, the assignor is not considered to have entered into a sale or a leasing transaction. If cash is involved in the sharing arrangement and the cash is pledged to the exploration and development of the property, no gain or loss should be recognized by the assignor, except if his share of the exploration and development costs under the contract are less than the cash received.

Shooting Rights
and Option Payments

As indicated in Chapter 2, before an oil and gas lease or other acquisition of mineral interests occurs, it is common for an oil and gas company to acquire "shooting rights" to conduct G&G exploration on land to determine its oil and gas potential. Shooting rights are sometimes used to avoid large lease bonus payments for areas that have questionable potential.

Shooting rights allow the holder exploration rights in a specified area. Any payments made to the assignor of the shooting rights are treated as G&G (see Chapter 5) by the oil and gas company. The assignor or landowner conveying the shooting rights treats the amount received as ordinary, nondepletable income.

Quite frequently, an oil and gas company will make a single payment that allows it to have shooting rights in an area, as well as an option to lease any or all of the land covered by the shooting rights. Normally, the entire amount paid is considered an option payment. Even though payments are normally stated in a per-acre amount, the entire amount paid for the option will probably be required to be capitalized as leasehold cost if the option is exercised on any portion of the acreage. If none of the acreage covered by the option is acquired, the cost of the option will be treated as an expense in the taxable year in which the option lapses.

As to the assignor or landowner, the payment received for the option is ordinary, nondepletable income in the year the option lapses. If a lease or leases are granted, the amount received for the option should be treated as part of the lease bonus, subject to cost depletion.

With respect to damages which may be paid by the oil and gas company to the landowner in connection with exploration activities, the landowner should be able to treat payments made specifically for damages to the surface as a return of capital and report no taxable income until the amount exceeds the tax basis in the property. If the landowner cannot prove that the payments are for damages to the surface and, hence, a return of capital, the payments should be reported as nondepletable, ordinary income. Payments for damages to crops should be reported as ordinary income.

Advance or Minimum Royalties

In some leasing transactions, the lessor will attempt to guarantee a minimum level of income by virtue of insisting on an advance or minimum royalty clause. Under such a provision, a payment is made based upon a hypothetical level of production during the term of the lease. Even if the hypothetical amount is not produced, the lessee pays the advance or minimum royalty amount to the lessor. In any event, depending upon the terminology included in the agreement, the lessee usually has a certain period of time to make up the advance payments from the lessor's share of production (referred to as a "recoupment" provision). If the advance royalty is not made up during the specified period, the lessee's payment does not count against later production. If the lessee's share of production exceeds the minimum royalties, of course, the lessee must pay the lessor the lessor's full share of production.

Advance and minimum royalties create a number of Federal income tax problems. The lessor must include the amount of payment as ordinary income in the year of receipt if using the cash method of accounting, or when the right to the payment becomes fixed if the lessor is an accrual method taxpayer. A lessor may claim cost depletion on advance or minimum royalties if the lessor has tax basis in the retained interest.

For many years, lessees had the option to deduct advance or minimum royalties in the year of payment, or in the year of recoupment. This option led to significant tax abuses in the 1970s. Since 1976, the deductibility of advance or minimum royalties has been subject to specific rules that limit the deductibility to the year paid or incurred.

The entire area of the tax treatment of advance or minimum royalties is complex, and care should be taken when these types of payments are involved to be sure that all parties have a clear understanding of the Federal income tax treatment to be accorded these payments.

Delay Rentals and
Other Payments to Lessor

In addition to advance or minimum royalty payments, other payment provisions are included in a typical oil and gas lease that guarantee the lessor a minimal amount of income before production occurs, and during interruptions in production.

The standard oil and gas lease provides that it will remain in effect for the primary term as long as drilling operations have commenced, or production is maintained in commercial quantities. If one of the two conditions is not met, the lessee may continue to lease during the primary term through the payment of delay rentals. Delay rental clauses may take different forms, but generally the delay rental payment is calculated on a per-acre basis and is paid annually to extend the term of a lease for one year. Failure to pay the delay rental terminates the lease.

The lessor treats delay rentals as ordinary income not subject to depletion. The delay rental payments may be deducted by the lessee as a current expense, or may be capitalized if the lessee so elects. Obviously, because delay rentals are currently deductible, a lessee will attempt to transform as much of the bonus payment into delay rentals as is possible. Such attempts, however, may result in treatment of the payments as an installment lease bonus, rather than delay rentals, for Federal income tax purposes.

Generally, oil and gas leases also contain a shut-in royalty payment provision that allows the lessee to maintain the lease when production ceases for a period of time. Shut-in royalty payments are treated as ordinary income by the lessor (not subject to depletion) and are deductible by the lessee as a current expense.

Production Payments Retained in Leasing Transactions

As indicated in Chapter 2, a production payment is an interest in minerals limited to a specific dollar amount, or a specific quantity of mineral extracted, which has a life shorter than the economic life of the reservoir. For example, an assignor may retain a production payment of $100,000 (plus interest) to be paid out of 50 percent of production, free of all costs and burdens. If the $100,000 (plus interest) will be paid within a period of time that ends before the end of the economic life of the reserves from which it was created, it should be a production payment for Federal income tax purposes. If not, it probably will be viewed as a continuing nonoperating interest in the nature of a royalty.

A production payment retained in a leasing transaction is treated as an economic interest in the mineral in place. Hence, the lessor realizes ordinary depletable income from payments under the production payment. The lessee treats the payment of the production payment as an installment bonus. Accordingly, the lessee must include the production used to satisfy the production payment in gross income and capitalize each payment (plus interest thereon) as part of the cost of the leasehold as paid or incurred. Although it would be simpler to include the entire amount of the production payment in the lessee's basis at the time of creation of the production payment, the IRS contends that each payment (plus interest) is capitalized only as paid or incurred.

Top Leasing

If a lease expires, leasehold cost can be written off by the lessee as a loss in the year of expiration (if the cost has not been previously written off as worthless) even if a new lease is subsequently executed for the property (assuming the new leasing transaction is independently negotiated).

In some cases, a "top lease," that is, a new lease between the same parties for all or part of the same property executed prior to the expiration of the old lease, is acquired. Since the rights of the lessor in that situation never terminate, no deduction will be allowed for the old leasehold cost, and any consideration paid for the "top lease" will also be capitalized as leasehold cost.

Acquisition Expenses

‌on of a fee simple or an oil and gas
‌s, including legal fees, broker-
‌g fees, and other costs may
‌ are capitalized and added
‌ acquired property. If more
‌d, these expenses should be
‌ properties so that a deduc-
‌ property to which some of
‌ abandoned.

CHAPTER 5

Geological and Geophysical Costs

General

As discussed in Chapter 2, before actual drilling is undertaken on a prospect, and in many cases before the prospect is actually acquired, a wide variety of G&G may be undertaken to determine the possible existence of oil and gas. Core drilling, gravity meters, seismic surveys, aerial photography from satellites, analyses of soil and rock and other techniques may be utilized in assessing the prospects of a particular location. These activities may be carried out by independent contractors or by an oil and gas company's own exploration personnel.

Historical Background
of Tax Treatment

For Federal income tax purposes, the proper treatment of G&G has been confusing for a number of years. Prior to 1941, the IRS and the oil and gas industry generally treated all G&G as ordinary and necessary business expenses deductible in the year paid or incurred. Since the 1940s, however, the IRS has treated G&G as capital expenditures related to the acquisition, or retention, of a leasehold.

G&G expenses lead to either further work on the property, or abandonment. While not entirely clear, it appears that an oil and gas company can take the position that G&G performed by its own staff should be ordinary and necessary business expenses incurred to locate and maintain an undeveloped lease inventory necessary for its continuation as a going concern. An argument can also be made that G&G performed prior to

the acquisition of properties are also ordinary and necessary business expenses. The IRS has treated G&G as capital expenditures for many years, and a taxpayer challenging that position can expect litigation.

Current IRS Position

An examination of the process in which G&G are undertaken is helpful in evaluating the current Federal income tax treatment of such costs. It is customary in the oil and gas business to conduct an exploration program in one or more identifiable project areas. Each project area encompasses a territory the oil and gas company determines can be evaluated advantageously in a single integrated operation.

The oil and gas company selects a specific project area to conduct a reconnaissance survey designed to yield data that will identify specific geological features with sufficient oil and gas producing potential to warrant further exploration.

Each separable, non-contiguous portion of the project area in which a specific geological feature is identified is a separate area of interest. The original project area is subdivided into as many small projects as there are areas of interest identified. When a detailed exploratory survey can be conducted without an initial reconnaissance survey, the project area and the area of interest are treated as being co-extensive by the IRS.

Detailed surveys are normally conducted on each area of interest. Data is sought to form a basis for making the decision to acquire or retain mineral rights within or adjacent to a particular area of interest, or to abandon the area of interest as unworthy of development.

It is the position of the IRS that G&G are performed for the purpose of obtaining and accumulating data to serve as a basis for the acquisition or retention of properties. Accordingly, these are capital expenditures and are not deductible as ordinary and necessary business expenses. Judicial precedents do not necessarily support such a broad conclusion on the part of the IRS, but taxpayers have not chosen to litigate the issue in recent years.

According to the IRS, if, on the basis of data obtained from preliminary work, only one area of interest is located and identified within the original project area, all G&G for the project area are to be allocated to the single area of interest. If two or more areas of interest are identified, the total costs are to be allocated equally among the various areas of interest. Interestingly, the IRS does not allow an allocation based on acreage in each area of interest even though that approach would seem to be a more appropriate allocation method.

If no areas of interest are located within the original project area, the entire amount of G&G attributable to the project area are deductible as a loss in the taxable year in which the project area is abandoned as a potential source of oil and gas production. If a property is acquired within an area of interest based on information obtained from a detailed survey, all G&G associated with the area of interest, when performed, are treated as capital costs of that property.

Where more than one property is acquired or retained within or adjacent to an area of interest, G&G allocable to each property are determined by allocating the entire G&G among the properties so acquired or retained on the basis of the comparative acreage of the properties. However, if no mineral rights are acquired or retained within or adjacent to an area of interest, the entire amount of G&G allocated to the area of interest are deductible as a loss in the taxable year in which such area of interest is abandoned.

From time to time, G&G associated with a specific area of interest are deferred until an identifiable event occurs which requires the costs to be assigned to specific properties, or justifies deducting those costs as a loss. Identifiable events include a lease sale that includes the area of interest involved, an indication that the area of interest is not available for lease, or an event establishing that the area of interest is worthless (e.g., test well in vicinity is a "dry hole").

Where exploration is conducted offshore, or on Government land onshore, the passage of 10 years without the area being made available for lease, or without an indication that the area is available for lease, is considered an event warranting deduction of the deferred costs as a loss. In the case of a non-government onshore lease, a passage of 5 years coupled with no indication that a lease is offered is deemed to be the occurrence of an identifiable event for the purpose of deducting the G&G as worthless. Note that these 5- and 10-year rules are merely guidelines promulgated by the IRS and other events could justify writing off the deferred costs at an earlier date.

Because the costs of a reconnaissance survey must be accumulated for the entire project and allocated equally to the areas of interest located within the project, oil and gas companies normally strive to limit the size of each project area upon which reconnaissance surveys are made, hoping to minimize the amount of G&G which must be capitalized. If project areas can be kept small, more areas are likely to be abandoned, generally resulting in less capitalization of G&G.

If a project area is very large, more than one survey

may be conducted on the property. If one or more of the surveys indicates no area of interest and the oil and gas company abandons efforts on that portion of the lease, the G&G attributable to that portion of the lease should be deductible at the time of abandonment. However, while a deduction may be taken for G&G allocable to that portion of the lease, the oil and gas company cannot take a partial abandonment loss for the leasehold cost attributable to that portion of the lease.

Bottom Hole and Dry Hole Contributions

Oil and gas companies from time to time enter into arrangements with adjoining or nearby lease owners whereby the other owners contribute cash to offset a portion of the cost incurred by the oil and gas company in drilling a test well on its property. The oil and gas company provides geological information about the well to those making contributions.

The first type of arrangement is referred to as a "bottom hole" contribution under which the contributing party agrees to make contributions regardless of whether the test well is productive or dry. The second type of arrangement is a "dry hole" contribution under which a contributing party makes a contribution if the well is dry. It is the position of the IRS that the contributing party must capitalize the contribution regardless of whether the test well is productive or nonproductive, presumably as G&G. However, if the well to which the contribution is made is dry, and the contributing party can prove that no value is added to its property by virtue of the expenditure, a deduction should be allowable for the amount of the contribution in the year the payment proves worthless.

The party receiving a "bottom hole" contribution is required to report the amount received as income and not as a reduction of IDC. Presumably, the recipient of a dry hole contribution should treat the amount received in the same manner.

Capitalization of Certain Costs

An active oil and gas company's exploration department generates numerous costs that do not result in the acquisition or retention of properties. Further, many exploration department activities do not relate to specific exploration projects, to specific areas of interest, or to specific properties. These costs should be deductible as ordinary and necessary business expenses, although the IRS has, from time to time, alleged that the costs must be allocated to activities that are identifiable to properties.

While far from clear at this time, it is possible that the IRS may apply the uniform capitalization rules of the Internal Revenue Code to exploration activities to require capitalization of general expenses not normally associated with exploration. The uniform capitalization rules do not appear applicable to most exploration and development costs, and a strong argument can be made that the uniform capitalization rules should not apply to G&G incurred in the exploration activity. However, the IRS has not clearly defined its position on this point.

Sharing Arrangements for Exploration and Development

General

The leasing and contractual arrangements described in Chapter 2 all relate to the sharing and shifting of risk among parties for the exploration and development of an oil and gas lease.

The most basic sharing arrangement is the oil and gas lease under which the lessee acquires the working or operating interest and bears all costs and responsibilities for exploration and development, leaving the landowner with a royalty as its share of the venture.

The bottom hole and dry hole contributions agreements described in Chapter 5 illustrate another basic form of sharing arrangement. Another common sharing arrangement involves the contribution of cash, goods, or services to the exploration and development of an oil and gas property in exchange for an interest in future production from the property.

From these simple illustrations, the types of sharing arrangements used in the oil and gas industry mushroom into a complex myriad of transactions. Before analyzing the various types of transactions and their tax consequences, it is first necessary to examine the pool of capital doctrine in more detail.

Pool of Capital Doctrine

The pool of capital doctrine is unique in oil and gas taxation and is perhaps one of the most controversial concepts in the Federal income tax law. Basically, the concept follows the industry practice of sharing risk and treats the exploration and development of an oil and gas property as a pooling of capital by the parties involved.

As a result, a contribution of property and services to the drilling of a well in exchange for a right to share in future production of that well does not result in the current recognition of taxable gain or loss by either the party contributing property or services or the recipient of those items. Taxpayers using a similar technique in a non-mineral transaction would normally have taxable gain from the exchange of property or services for an interest in future revenues.

The pool of capital doctrine has its foundation in a U.S. Supreme Court decision defining "economic interest." The doctrine developed from the concept that a taxpayer claiming depletion must have an economic interest in the property. For many years, industry practice, court cases, and the IRS recognized the theory and did not treat a contribution to the exploration or development of an oil and gas property as a taxable event.

Recently, the IRS has attempted to curtail, and perhaps eliminate, the application of the doctrine in the oil and gas industry. Nevertheless, the doctrine continues to be used in many transactions. Taxpayers must recognize, however, that some limitations have been developed by the IRS and the courts over the years, as discussed more fully below.

Farmouts and Related Transactions

The transaction that most clearly demonstrates the applicability of the pool of capital doctrine is the "farmout" arrangement under which the owner of a working interest transfers, or farms out, all or a portion of the working interest in a drillsite, and, in many cases, surrounding acreage, to another party in exchange for that party's agreement to drill a well on the drillsite. The transferred interest is typically retained by the assignee until the assignee obtains complete payout of its investment in the well. After payout, a portion of the transferred interest generally reverts to the original owner and subsequent production is allocated as agreed between the original owner and the other party for the balance of the life of the property. In these transactions, the party transferring the working interest is referred to as the "farmor," and the party receiving the transfer is referred to as the "farmee."

The farmout is simply a form of sharing arrangement under which the working interest owner assigns all or part of the working interest (as contrasted to assigning a nonoperating interest) to another party in exchange for that party's assumption of all or part of the exploration and development burden.

A less traditional form of farmout involves the assignment of a nonoperating interest. While the transaction is a sharing arrangement to which the pool of capital doctrine applies, it does not fit the traditional farmout pattern. In this transaction, an oil and gas company retains the working interest, but transfers a nonoperating interest, such as a royalty or net profits interest, to another party in exchange for that party's obligation to drill a well. In that case, the working interest owner retains its entire tax basis in the lease as depletable cost, and the assignee of the nonoperating interest capitalizes all costs of drilling and equipping the well as the depletable cost of the nonoperating interest received. The assignee has no capitalized equipment cost and no IDC deduction.

For Federal income tax purposes the farmor in a traditional farmout arrangement has no gain or loss and treats its entire tax basis in the working interest as the tax basis of the fractional working interest retained. The farmee must capitalize the portion of the IDC and equipment cost applicable to the portion of the working interest retained by the farmor as leasehold cost. The portion of the IDC and equipment cost incurred for the farmee's fractional share of the working interest would be subject to the general rules regarding IDC expenses and the capitalization of equipment costs.

If, in a farmout transaction, the farmor retains an overriding royalty interest or net profits interest in the property, the farmee owns all of the working interest and should be allowed to deduct all IDC and depreciate all equipment costs incurred for the working interest. The farmor merely transfers its tax basis in the working interest to tax basis in the retained nonoperating interest. If the retained nonoperating interest is later converted back into a working interest, the timing of the conversion may have a significant Federal income tax effect on the farmee. If the nonoperating interest cannot be converted back into a working interest until "payout," or thereafter, no Federal income tax problem should result to the farmee since the farmee owns all of the working interest during the payout. If the nonoperating interest is convertible prior to payout, the transaction will nevertheless be treated as a sharing arrangement with neither party recognizing gain nor loss from the conversion. If, however, conversion can occur prior to payout, the farmee may not own all of the working interest for the complete payout period and must capitalize the portion of the IDC and equipment costs applicable to the fractional interest subject to pre-payout conversion as leasehold cost.

Even if the conversion occurs at or after payout, the farmee is required to transfer the depletable and depreciable basis attributable to the portion of the working interest surrendered to the depletable basis of the portion of the working interest the farmee continues to own.

It is quite common in farmout arrangements for the farmor to assign all of the working interest in the drillsite, and a different fraction in surrounding acreage, in return for the farmee's obligation to drill and equip one or more wells. Historically, these types of arrangements have been treated as nontaxable under the pool of capital doctrine. In 1977, however, the IRS limited the application of the pool of capital doctrine in these transactions, holding that the doctrine applies only to the arrangement relating to the drillsite and not to the arrangement relating to the assignment of the surrounding acreage.

For example, assume that an oil and gas company owns a property consisting of 640 acres and that a drillsite of 160 acres has been identified. Assume further that a farmout agreement is entered into with another party providing that, upon the abandonment or completion of a well on the 160-acre drillsite, the farmor assigns the entire working interest in the drillsite to the farmee, subject to a retained overriding royalty interest, and also assign to the farmee an undivided one-half interest in the working interest in the remaining 480 acres. The agreement further provides that the retained overriding royalty interest can be converted into an undivided one-half interest in the drillsite working interest after payout. In that case, the IRS takes the position that the assignment relating to the drillsite is nontaxable under the pool of capital doctrine, but that farmee has taxable income from the assignment of the undivided one-half interest in the working interest in the remaining acreage equal to the fair market value of the interest received in the remaining acreage. Presumably, the farmout divides the 640-acre tract into two separate tax properties, one being the working interest in the drillsite and the other being the undivided one-half interest in the working interest in the remaining acreage. Further, the oil and gas company making the assignment of the undivided one-half interest in the surrounding acreage is considered to have made a sale of the interest and is required to report the difference between the fair market value of that interest and the tax basis allocable to that interest as gain or loss in the year of assignment.

The position of the IRS has dramatically affected the form of farmout transactions in the oil and gas industry. One approach used to avoid the problem with assignments of acreage outside the drillsite is to make the entire assignment of both the drillsite and the surrounding acreage before drilling commences to attempt to keep the value of the surrounding acreage as low as possible at the time of assignment. Another approach is to require payout on the entire tract, rather than on a well-by-well basis. In addition, taxpayers in the oil and gas industry frequently use tax partnerships to own farmout arrangements to avoid the consequences of the IRS's position regarding the acreage outside the drillsite.

In some cases, one of the parties to a sharing arrangement transfers cash or other consideration in addition to the work performed, or in addition to the property assigned. If such cash or other property is pledged and used in development of the property, no taxable income should result. The recipient of the funds should be allowed to credit the cash or the value of other consideration received against his costs incurred for the property.

Services for an Interest
in an Oil and Gas Property

For many years, geologists, drillers, suppliers, attorneys, brokers, and accountants have provided services in sharing arrangements during the exploration and development activities relating to a property and have received an interest in the property for such services. Under normal Federal income tax rules, property received for services is taxable at its fair market value at the time of receipt, unless certain restrictions on the ownership of the property exist at the time the property is received. However, as previously discussed, because of the pool of capital doctrine, it has been common practice in the oil and gas industry to treat the services as a contribution to the pool of capital, and the receipt of the oil and gas property interest as non-taxable.

Both the IRS and the industry utilized the pool of capital doctrine for many years as the rationale for treating the receipt of an oil and gas property interest for services as non-taxable. Presumably, the difficulty in valuing such interests, particularly interests in non-producing properties, led to this administrative and industry practice.

More recently, there have been several court decisions and administrative rulings by the IRS which seek to limit the applicability of the pool of capital doctrine in cases where a property interest is received for services. Further, general provisions of the Internal Revenue Code have been applied to attempt to limit the applicability of the pool of capital doctrine in certain cases.

It now appears that where a property interest is received by an employee in lieu of cash compensation, and in situations where the interest is received as a result of a contractual obligation to perform the services, the IRS likely will attempt to tax the value of the property interest at the time of receipt. If the property interest received is non-transferable or subject to a substantial risk of forfeiture, the general provisions of the Internal Revenue Code do not tax the recipient on the property interest's value until (1) the lapse or removal of the substantial risk of forfeiture, or (2) the property interest becomes transferable. The recipient of the property interest can elect to report the income upon the receipt of the property interest, despite the existence of a substantial risk of forfeiture. Obviously, if an interest in an undeveloped property is received, and it is believed that the pool of capital doctrine will not be available to treat the receipt as tax-free, recipients of a property interest for services should consider electing to report the value of the interest at the time received. If an election is made to have the property taxed at the time received and the property is later forfeited, the taxpayer will not be allowed a loss deduction for the amount included in the income under the election.

While relatively unexplored at this time, it appears that employees and independent contractors desiring an interest in future revenues from an oil and gas property for services rendered should consider the use of an unsecured, unfunded contractual promise to pay, rather than a real property interest. It appears that an unsecured, unfunded contractual interest does not give rise to current taxable income under the general provisions of the Internal Revenue Code and that the recipient will be taxed only on income if, as, and when received in the future. Further litigation in this area can be expected.

C H A P T E R 7

Carried
Interests

General

A "carried interest" is an agreement between two or more working interest owners under which one working interest owner is obligated to pay all or part of the exploration and development costs attributable to the working interest held by others and recover those amounts out of future production (if any) from the other party's share of the working interest. A working interest owner whose costs are paid by another person is referred to as the "carried party" since another party is paying, or carrying, that working interest owner's share of the exploration or development obligation. The party paying another party's obligation is referred to as the "carrying party." A carried interest normally arises from a farmout transaction in which all or part of the working interest is assigned to a party who pays all or part of the exploration or development expenses as described in Chapter 6.

In some carried interest arrangements, the period of the carried interest may last for the entire economic life of the property. In others, it is limited to the drilling of one well, or a specified number of wells. A carried interest for the full economic life of the property (often referred to as an "unlimited carry") is treated as a net profits interest by the farmor for Federal income tax purposes.

Types of Carried Interest

Carried interests have historically taken at least three different forms, but are all currently treated in the same manner by the IRS.

The first is referred to as an Abercrombie type of carried interest. In an Abercrombie type, the working interest owner assigns a portion of the working interest to an assignee who is obligated to drill and equip a well. The assignee pays all the costs of drilling, equipping and operating the property. The working interest owner also gives the assignee a lien on the interest retained by the working interest owner to assure recovery of costs by the assignee. After the assignee recovers all costs (that is, achieves payout), the parties share revenue and expenses on an agreed upon basis. In the Abercrombie form of carried interest, the carrying party effectively owns all the working interest until payout and is allowed to deduct IDC and to capitalize and depreciate equipment costs.

The second type is a Herndon carried interest. In the Herndon situation, the working interest owner assigns a portion of the working interest to an assignee. Under the terms of the assignment, the assignee drills and equips a well and pays all the costs. In addition, the working interest owner assigns a production payment to the assignee payable out of the working interest owner's share of revenue until the assignee achieves payment. Early court decisions allowed the assignee to deduct IDC and equipment depreciation attributable to its fractional interest of the working interest, but the remaining costs had to be capitalized as the cost of the production payment. However, it now appears that the IRS will look to the economic substance of the Herndon type arrangement and treat the assignee as owning the entire working interest during the payout period. Hence, the assignee should be allowed to deduct the IDC and capitalize and depreciate equipment costs.

The third type of carried interest is the Manahan. In that situation, the working interest owner assigns the full working interest to an assignee who drills and completes a well on the property. At the time of complete payout, a portion of the working interest reverts to the working interest owner. Since the assignee actually owns all of the working interest during the payout period, the assignee is entitled to all income and deductions applicable to working interest during that time, including IDC and depreciation on equipment.

As indicated above, it appears that the IRS will now treat Abercrombie, Herndon, and Manahan carried interests identically for Federal income tax purposes. Hence, the carrying party should be entitled to deduct IDC attributable to its share of the working interest as well as capitalize and depreciate the cost of equipment attributable to such interest.

In all types of carried interest arrangements, the carrying party must, at conversion by the carried party, transfer the depletable and depreciable basis attributable to the portion of the working interest surrendered to the leasehold costs of its continuing interest.

CHAPTER 8

Unitizations and Poolings

General

Unitization or pooling agreements are common forms of sharing arrangements. Most oil and gas leases contain a pooling clause which allows the assembling of acreage to create a drilling unit in accordance with the spacing requirements in the jurisdiction in which the well is drilled. Poolings may be voluntary, or in some states, may be required by the state regulatory agency.

The terms *pooling* and *unitization* are sometimes used interchangeably. However, a unitization is normally more than the mere accumulation of acreage to create a drilling unit. Generally, a unitization combines the economic interests in a reservoir to permit the joint exploration, development and operation in a more efficient manner. Unitization agreements provide for operation on a joint basis and set forth the method in which production from the unitized properties is shared by the parties.

Unitizations require substantial engineering and legal expenses. Those expenses incurred in a state-mandated unitization are deductible as ordinary necessary business expenses. Presumably expenses incurred in voluntary unitizations receive the same treatment.

Unitization Agreements

A unitization agreement may provide for cross-assignments of ownership interests, but no formal cross-assignments are necessary to have unitization treated as a tax-free exchange of like kind assets for Fed-

"Good work, dear."

eral income tax purposes. An oil and gas company contributing several working interests to a unit will have one tax property in the unit for Federal income tax purposes.

If only a part of a property is included, the total tax basis of the property on the date of unitization is allocated based upon the relative fair market values of the two parts of the property. If a portion of a tax property becomes part of a unit, two separate properties are created for most Federal income tax purposes.

If the unit is terminated, the tax basis of the unit is folded into the non-unitized portion of the property. If the non-unitized portion of the property is abandoned, while the unitized portion continues to be held, no abandonment loss is recognized since the two portions are treated as one property for purposes of determining the proper time for an abandonment loss for Federal income tax purposes.

Equalizations

In determining the total value of a unit, prior expenditures and future development costs are considered. Each contributing party's share of the value is determined by multiplying that party's participation factor by the total value of the unit. It is common for differences to exist between the value of the share of the unit received and the value of the contribution made. Hence, "equalizations" are required. Equalizations may be in the form of (1) cash, (2) adjustments to future revenue sharing, or (3) adjustments of future expenditures.

Equalizations made in cash are paid to those parties who contribute value in excess of the value of their share of

A very independent producer.

the unit. Those receiving value in the unit in excess of their contribution pay the cash equalization. Equalization payments may be in a lump sum, or separate payments may be made for equipment and IDC. Since a unitization is treated as a tax-free exchange for Federal income tax purposes, cash equalization payments are treated as "boot." Losses are not recognized, but gain is recognized to the extent of boot received. The nature of the gain depends upon whether the boot relates to IDC or equipment. If separate equalization payments are made for equipment and IDC, the calculation of the gain resulting from the two types of assets should be relatively simple. However, lump sum equalization payments require an allocation between equipment and IDC on the basis of relative fair market values. Because of the complexity of lump sum equalization payments, it is now common for most unitization agreements to provide for separate equalization of IDC and equipment. The parties should attempt to use this approach to avoid Federal income tax complications.

As indicated above, equalization may also occur through either adjustments of participation in future revenues, or adjustments of subsequent expenditures. Assuming the adjustment is made by merely adjusting future revenue participation, and no other property interests are involved in the transaction, no gain or loss should be recognized by the parties. If the adjustment is made through adjustment of future expenditures and a party pays more than its proportionate share of post-unitization costs, it appears the party bearing the additional cost is deemed to have incurred additional leasehold cost. The portion of the future expenditures attributable to the paying party's own interest in the unit, however, should be treated as IDC applicable to the paying party's interest in the unit or as equipment applicable to that share of the unit.

C H A P T E R 9

Gross
Income

General

The gross income from an oil and gas property is the amount for which the taxpayer sells the oil and gas in the immediate vicinity of the well. This amount is essential in computing percentage depletion (see Chapter 12). If production is not sold in the immediate vicinity of the well, the gross income is calculated based on the representative market or field price of the oil and gas before conversion or transportation. Conversion is any manufacturing process by which the production is converted into a semi-refined or refined product. Transportation is the carrying of production from the well to a point where refining or marketing facilities are located.

The determination of gross income from the property may be made easily if there is an actual sale of the production or there is a current posted market or field price. Absent such information, however, the price must be "backed into" by adjusting the marketing location price to determine the price of the oil or gas before transportation and before conversion. If readily identifiable, the transportation cost or conversion cost may be simply subtracted from the selling price at the refining or marketing facility to determine the gross income from the property for percentage depletion purposes.

Production vs. Manufacturing

One of the most difficult areas in oil and gas tax law is the determination of whether a procedure applied to oil or gas is a production or manufacturing process. Frequently, production from a well is taken through a gravity separator on the lease to remove water from the

Different perspectives on the same transaction.
(I.R.S. Tax Audit)

oil and to permit some separation of lighter and heavier hydrocarbons. The gravity separator is a production process and not a refining process. Hence, the proceeds from the sale of the separated product need not be adjusted to determine the sales price. While the IRS does not always agree, separation on the lease, regardless of the type of separator utilized, should be treated as a production process so long as the purpose is merely to break down the component parts of the hydrocarbons produced from the well.

Gas plants and recycling plants normally involve producing and manufacturing processes. If there is a posted price for oil or gas at the wellhead, the portion of the sale proceeds of the plant products equal to the posted price probably represents the gross income from the property for percentage depletion purposes. If posted prices are not available, however, other means of determining gross income of the property must be used.

While it appears that the separation process should, in most cases, be a production function, the issue is not as clear-cut with respect to the absorption phase and the fractionation phase. For example, compression of separated gas to meet specifications of the purchaser constitutes manufacturing. Hence, the value added by compression does not constitute gross income for the property for percentage depletion purposes. The IRS takes the position that *any* step beyond separation constitutes manufacturing.

Court cases have generally agreed that absorption and fractionation are manufacturing functions. With respect to recycling plants, however, the absorption process in such plants has been determined to constitute a production function. In a recycling plant, when the gas and oil streams are brought back into contact after the separation process, additional hydrocarbons are separated from the gas (the absorption process). After that separation, the residue gas is repressured and reinjected into the reservoir. In that case the absorption function to separate the residue gas should be treated as part of the production process. While the IRS does not appear to agree, the most logical method of computing gross income from the property is to determine the value of the oil or gas when it first becomes commercially marketable. At that point, gross income from the property should be computed and any remaining gross income realized from further processing would not be treated as gross income from the property, but rather would be classified as manufacturing income.

Determining gross income from the property for percentage depletion purposes where significant processing of production occurs remains unclear. Accordingly, care should be taken in computing the amount to be reported as gross income from the property where processes beyond the separation process are involved in preparing the oil or gas for sale.

Take-or-Pay Contracts

If an oil and gas company enters into a long term gas sales contract, that contract may contain a "take-or-pay clause," under which the buyer will pay for quantities of gas whether or not the gas is delivered to the buyer. In the normal situation, the buyer is allowed to credit payments made for gas not taken against future deliveries under the contract.

The IRS has ruled that payments received for gas not taken under a take-or-pay contract are gross income to the oil and gas company, even though these payments are similar to a production payment (under which the oil and gas company is not viewed as receiving gross income). At least one court has concurred in the IRS's view.

To avoid this result, a taxpayer receiving payments

under a take-or-pay contract should consider adopting a method of accounting for those payments which allows deferring income recognition. Care should be taken that the election is made properly and is appropriate for the taxpayer's facts and circumstances.

The buyer under the take-or-pay contract will treat payments made under the contract as deferred charges, if it is believed that the gas paid for, but not taken, is recoverable from future production. As the gas is taken in future years, the deferred charges would be adjusted so that the cost of the gas is charged against taxable income in the year in which quantities are made up. If the gas is not taken and it can be proven that the amounts paid will not be refunded, the buyer should be allowed a loss for the deferred charges in the year it is determined that such charges are worthless.

Gas Advances

It is common for parties needing a secure gas supply to advance funds for the exploration and development of the properties from which the gas is to be delivered. Normally, the party advancing the funds is granted the right to purchase all or a part of the gas produced from the properties for which the advance is made. Advances normally take the form of non-interest bearing (and usually nonrecourse) notes.

Recipients of advances for gas exploration and development have traditionally treated the amounts as an interest-free loan with no tax consequences. The IRS has, from time to time, asserted that an advance represents proceeds from a sale of royalties, and not a loan, if the taxpayer cannot prove that the gas advance will be repaid prior to the end of the economic life of the properties to which the advance relates. The IRS has also asserted that gas advances constitute gross income to the recipient at the time the advance was made.

Since 1984, the Internal Revenue Code has provided for the treatment of non-interest bearing, or low interest, loans. Under the present rules in the Internal Revenue Code, the party advancing the funds is deemed to have imputed interest income from the transaction and the party receiving the advance is deemed to have imputed interest expense.

Because of the continuing uncertainty surrounding gas advances, care should be taken in these transactions to be sure that the gas advance is a loan, and not a sales transaction, to avoid unexpected tax consequences.

Oil and Gas Exchanges

Quite often oil and gas companies exchange oil or gas to save transportation charges or for other reasons. If different qualities of oil or gas are involved in the exchange, adjustments are made for the differences.

If an oil and gas company receives oil or gas from another party and is obligated to return oil or gas in kind, the receipt of the oil and gas should not be viewed as a taxable transaction. The sale of the oil or gas received should be treated as a short sale (similar to treatment of a short sale of securities) and no gain or loss should be reported at the time of the sale. Upon repaying the oil or gas borrowed, the short sale is deemed closed, and ordinary gain or loss is recognized at that time by the party delivering the oil or gas in the repayment of the amount borrowed. Unfortunately the

IRS has held, contrary to the above, that an exchange of production is taxable at the time of the exchange.

Gas is often borrowed to meet short-term needs with an agreement to repay with a greater quantity of gas. Presumably, the borrowing party could treat the transaction as a sale, measuring its gross income from the transaction as the estimated value of the gas to be repaid. Another possible treatment would be as a loan between the parties.

Gas Balancing Arrangements

The IRS and the industry have long considered the proper Federal income tax treatment of a gas balancing arrangement, which governs production and gas delivery from a property and resolves imbalances among the owners of the property.

To analyze the gas balancing agreement, it must be determined whether each amount of gas produced belongs jointly to the owners, or whether the owners can partition the ownership of the gas in the ground. If the position is taken that a partition has occurred, the taxpayer receiving the sales proceeds would record gross income in accordance with its method of accounting, regardless of its percentage of ownership in production. For example, a 25 percent owner who takes 50 percent of the gas production would record the gross income from that 50 percent of production so long as the remaining reserves in the ground equal, or are greater than, the cumulative imbalance. No liabilities or receivables are recorded by the parties. Since the party taking the production does not reflect all of the expenses attributable to the production sold, a proper matching of revenues and expenses does not occur under this method.

The other approach to gas balancing contracts is to have the parties report only those shares of gross income allocable to their ownership interest in the producing property. Using the foregoing example, a 25 percent owner would record gross income of 25 percent even though that owner takes 50 percent of the gas production. Under this approach, a better matching of revenues and expenses occurs for Federal income tax purposes.

The IRS has provided some guidance for gas balancing contracts, but many questions remain.

Intangible Drilling and Development Costs and Depreciable Equipment

General — IDC

As indicated in Chapter 1, the IDC deduction is allowed to provide a mechanism to attract capital for the high risk business of exploring for, and developing, oil and gas. While the IDC deduction remains available, its value as a capital formation technique has been diminished by "back door" limitations and penalties, such as the AMT and the at-risk rule (see Chapter 14). Nevertheless, the IDC deduction remains an important, albeit limited, incentive for oil and gas exploration and development.

After G&G are completed and a potential drillsite has been located, preparation for exploration of the property commences. As discussed in Chapter 2, sharing arrangements are often created at this point for the exploration or development of a property. The party holding the working interest is responsible for drilling and other costs inherent in the exploration and development activity.

Numerous expenses are incurred in preparing a property for drilling and production, including final G&G at the drillsite (which should qualify as IDC, rather than be treated as G&G), road construction, site clearing, and installation of the drilling rig. Since all of these costs are normally permanent improvements under general Federal income tax rules, these costs would normally be capitalized and recovered through depletion or depreciation. However, because of the IDC election, the party incurring the costs may elect to currently deduct them for Federal income tax purposes.

If the election to expense IDC is not properly made, the party incurring the IDC is required to capitalize costs as either equipment or leasehold cost. If a party does not elect to deduct IDC, a second election is available to

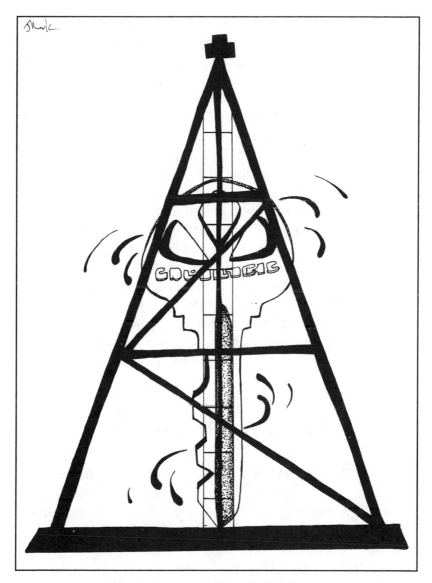

TURN-KEY DRILLING

capitalize or expense IDC attributable to dry, or non-productive, wells.

The election to expense IDC must be made for the first tax year in which IDC are incurred. Once the election is made, it is binding for all subsequent years. The election regarding the treatment of the costs of dry holes does not have to be made until the first year in which a dry hole is drilled. If the election is made to deduct dry hole costs, those costs are allowed as a loss in the year the dry hole is completed and abandoned. Of course, a taxpayer electing to expense IDC or dry holes will not be able to include those costs in the tax basis of the property upon which such costs are incurred.

As discussed in Chapter 14, taxpayers should be aware that IDC incurred on productive wells may be subject to unfavorable treatment under the AMT, the passive loss limitation and the at-risk rule. Further, taxpayers electing to expense IDC in the year paid or incurred should realize that all or a portion of the IDC may have to be recaptured as ordinary income upon the sale or disposition of the property as discussed below.

IDC as Tax Preference

In 1976, IDC became a tax preference item, first for minimum tax purposes, and now for AMT purposes. Recent legislation has softened the effects of this limitation by limiting the effect of this preference item for IDC incurred by independent producers and royalty owners (the "IPRO" exception) subsequent to 1992 (For a further discussion of the IPRO exception, see Chapter 12).

Because of the complications created by the 1976 change, taxpayers who had not previously done so elected to capitalize IDC and deduct the capitalized amount ratably over a 10-year period to avoid having such IDC taken into account for AMT purposes. It appears that a taxpayer may elect to capitalize some IDC and expense other IDC on the same property. After the election is made for a particular expenditure, however, it can only be revoked with the consent of the IRS. The amounts capitalized under this special election are included in the tax basis of the oil and gas property for purposes of determining gain or loss on the sale or disposition of the property.

The recent change in the law regarding IDC as a tax preference item diminishes the attractiveness of the 10-year amortization period.

Integrated Oil and Gas Companies

Special rules relate to the IDC incurred by integrated companies (i.e., companies which not only produce oil and gas, but also refine or market oil and gas products). After 1986, integrated oil companies must capitalize 30 percent of the IDC which would otherwise be deductible under their IDC election. These capitalized costs may be deducted ratably over a 60-month period beginning in the month in which the costs are paid or incurred.

The Payout Concept

Only the owner of the working, or operating, interest in a property has the responsibility for exploring and developing the property and incurring IDC. Hence, the election to expense IDC is made by the working interest owner when IDC are first paid or incurred. As discussed in Chapters 6 and 7, the IRS allows a taxpayer to deduct IDC only to the extent attributable to its share of the total working, or operating, interest in the well. IDC attributable to working interests held by others must be capitalized as leasehold cost.

In order to deduct the IDC attributable to a taxpayer's share of the working interest, the taxpayer must hold that share of the working or operating interest until complete payout. For example, if a taxpayer agrees to drill and equip a well on a lease in return for a 50 percent interest in a lease, the taxpayer is required to capitalize 50 percent of the IDC as depletable cost and is allowed to deduct the other 50 percent as IDC. On the other hand, if the party earning the interest in the lease for drilling the well is entitled to 100 percent of the working interest income and is required to pay all expenses until net proceeds from the working interest equal that party's total costs, the drilling party is entitled to deduct all the IDC and capitalize and depreciate all of the tangible equipment costs. If, at the time of payout, 50 percent of the lease reverts to the original owner, 50 percent of the remaining tax basis in the equipment is transferred to the depletable basis of the working interest retained.

If an oil and gas company can look to the aggregate gross income from two or more separate tax properties to recoup the total cost of wells drilled on each of the properties, the IRS takes the position that the taxpayer's deduction for IDC is limited to the portion of the IDC attributable to the taxpayer's permanent working or operating interest in each tract. For example, assume a taxpayer agrees to drill and operate a well on each of two non-contiguous tracts in exchange for the entire working interest in each tract until the aggregate net proceeds from both properties equal the total costs incurred by the taxpayer for the two wells. Further assume that the taxpayer will have a 75 percent permanent operating interest in each tract after payout. Under the IRS's position, the taxpayer's deduction for IDC is limited to 75 percent, and the remaining IDC is capitalized as a leasehold cost. The rationale for the IRS's position is that if the net proceeds from one property are sufficient to pay the total costs on both properties, it would be possible for the taxpayer to reduce his interest to the 75 percent permanent working interest level in the two tracts without having held the entire working interest in both tracts during the complete payout period for each individual tract. To avoid this problem, simply compute payout separately for each tract.

Making the
IDC Election

As indicated above, a taxpayer is required to make a clear election to deduct IDC. Without the election, a taxpayer is required to capitalize IDC. To avoid any controversy with the IRS, the election should be made by deducting IDC in the taxpayer's Federal income tax return for the first year in which IDC are incurred and by specifically stating in writing that an election to deduct IDC is being made in the return. While not entirely clear, it appears that an election to deduct IDC made in a delinquent return may not be considered timely made and the taxpayer may have permanently lost his ability to deduct IDC.

Timing of
IDC Deduction

If a proper election to deduct IDC has been made, the timing of the IDC deduction is determined by the taxpayer's method of accounting.

As a general rule, taxpayers using the cash method of accounting claim the IDC deduction in the taxable year in which it is actually paid. On the other hand, taxpayers using the accrual method of accounting generally must take the deduction in the taxable year in which all events occur fixing the liability for the IDC and the year in which economic performance occurs. The general principles, however, are sometimes difficult to apply to transactions used in the oil and gas industry. In addition, the IRS rules relating to accrual basis taxpayers are very complex and often are difficult to apply.

For many years, a serious issue involved whether prepaid IDC could be deducted in the year paid, even if substantial services had not been performed by the drilling contractor at the time of prepayment. If a substantial amount of IDC is prepaid, a material distortion of taxable income can result from a large deduction in the year of prepayment. One can argue that prepaid IDC creates an asset having a useful life extending beyond the end of the year of prepayment. While that prepayment of IDC for valid business reasons should be deductible in the year paid, the IRS will certainly review such a transaction carefully and may attempt to disallow the deduction in the year paid on the basis that it creates a material distortion of income.

For an accrual basis taxpayer to deduct IDC, all events which fix the amount of the liability and economic performance must occur in the taxable year for which the deduction is claimed. Economic performance is deemed to occur when the goods and services are provided. There is an exception to the economic performance test for certain recurring items which presumably would include IDC for an oil and gas company actively engaged in exploration and development. If the technical requirements of the exception can be met by the taxpayer, it appears that accrued IDC can be deducted in the year the liability becomes fixed.

After 1986, "tax shelters" are prohibited from using the cash method of accounting. Further, a tax shelter using the accrual method of accounting may not use the exception to the economic performance rule mentioned above. As a result, under the current tax rules, a deduction for prepaid IDC, whether incurred by a cash or an accrual basis taxpayer, may be difficult to sustain in the year of prepayment.

Identifying IDC

Identifying IDC which are properly deductible under the IDC election requires a careful review of the various elements of drilling activity. IDC generally includes any cost incurred that in itself has no salvage value and is necessary for the drilling of wells, or the preparation of wells for production of oil and gas.

IDC include wages, fuel, repairs, supplies, and similar items which are used in (1) drilling, shooting, or cleaning wells, (2) clearing ground, draining, road making, and surveying, as well as G&G to prepare the drill site, and (3) construction of rigs, tanks, pipelines, and other physical structures necessary for the drilling of wells and the preparation of wells for the production of oil and gas.

The cost of installing tangible equipment placed in the well itself is included as IDC although the cost of the equipment must be capitalized and depreciated. Operating expenses required for the production of oil and gas, as well as installation charges for equipment, facilities and structures not incident to, or necessary for, the drilling of wells are not treated as IDC.

Care should be taken to distinguish between installation costs that are treated as IDC and those that are capitalized and recovered through depreciation. Normally, installation of equipment necessary for drilling wells and preparation of wells for production are treated as IDC. The IRS regards a well as completed when the casing and Christmas tree have been installed. Installation costs to the point of completion are considered IDC. Installation costs incurred for pumping equipment, flow lines, separators, storage tanks, treating equipment, salt water disposal equipment, and similar items which are treated as production facilities must be capitalized and depreciated.

The cost of surface casing cemented in place has been held not to be IDC. The costs of this casing must be capitalized and depreciated. While this treatment of surface casing is very questionable, taxpayers attempting to deduct the cost of such casing as IDC may be challenged by the IRS. Labor and other costs expended in installing the surface casing and cementing it in place are treated as IDC.

Offshore IDC

Because of the substantial activity in offshore drilling, numerous questions have been raised regarding the treatment of various costs incurred in connection with such activities. As discussed in Chapter 2, to evaluate an offshore prospect, exploratory wells are normally drilled from mobile rigs to determine the existence and extent of deposits. These wells are usually plugged and abandoned, instead of being completed as production wells. After sufficient quantities of oil and gas are determined to be present, a drilling platform is constructed for the development of the property.

Obviously, the cost of the mobile drilling rig is capital and must be depreciated. However, costs incurred for wells which are drilled without an intent to produce oil and gas are deductible as IDC. If the well is meant to be an offshore stratigraphic test well in which the taxpayer has no economic interest, however, the amounts which would normally constitute IDC for the well must be capitalized as G&G.

The IRS has ruled that the deductibility of the costs of constructing and installing offshore drilling platforms as IDC must be determined on a case-by-case basis. On-shore design and fabrication expenditures may be deducted as IDC if evidence demonstrates that (1) the platform is necessary for drilling wells, (2) the platform is designed and constructed for use at a specific site, and (3) platforms of that type are not ordinarily reused or otherwise salvaged as a unit. Once it is determined that the platform cannot be salvaged, the same analysis is applied to each of the platform's structural components and sub-components. Only those components designed for permanent use at a specific site may be deducted as IDC. The IRS specifically treats the cost of purchased items and the cost of fabricating the drilling equipment and machinery to be installed on the platform as capital items subject to depreciation.

Service Wells and Other Special Cases

Costs in the nature of IDC may be incurred in connection with the various activities associated with producing oil and gas which may or may not be treated as IDC for Federal income tax purposes. The following illustrates several of the situations where such costs may be incurred:

1. Quite often wells other than oil and gas wells are drilled in exploration, development, and production activities. Such wells are known as "service" wells. If a service well is incident to the drilling of an oil and gas well, or the preparation of a well for production, and does not have a salvage value, the cost of the service well is classified as IDC. If the purpose of the well is to furnish water for drilling operations, the water well is considered incident to, and necessary for, drilling the oil and gas well. Hence, its cost will qualify as IDC.

2. The cost of drilling an input well to assist in producing oil or gas is treated as IDC. While it is not clear that the IRS views input wells necessary for secondary and tertiary production in the same manner, such costs should be treated as IDC also.

3. The acquisition and use of subsurface natural gas storage facilities, including drilling of injection and withdrawal wells, are capital costs which are depreciable and are not deductible as IDC. Further, the cost of drilling an aquifer for use as a gas storage reservoir is treated as depreciable, not as IDC.

4. Salt water disposal wells are not incident to and necessary for preparing wells for production.

Hence, the cost of a salt water disposal system must be capitalized and depreciated.

5. If a water supply well is drilled to obtain water for use in a secondary or tertiary recovery process, it is not a well drilled to prepare a well for production. As a result, the cost of a water supply well is capitalized and depreciated. The same treatment is accorded a well drilled to obtain carbon dioxide for tertiary recovery processes.

6. After a well commences production, sand and other problems may be encountered which require the well to be shut down for workover. Normally, tubing is pulled and the inside of the casing is washed out with mud. In addition, explosives may be set off at the bottom of the well to dislodge accumulated silt and sand. These costs are operating costs and should be treated as ordinary and necessary business expenses. If a well is deepened in the process, the cost attributable to the deepening of the well should be treated as IDC.

Recapture of IDC and Depletion

Upon the disposition of an oil and gas property at a gain, IDC or depletion (see Chapter 12) may be recaptured as ordinary income. For properties placed in service prior to 1987, the amount subject to recapture is the lesser of (1) the IDC previously deducted (subject to certain adjustments) or (2) the gain realized. For properties placed in service after 1986, the amount subject to recapture as ordinary income is the lesser of (1) the IDC and depletion (but not percentage depletion in excess of the property's tax basis) previously deducted or (2) the gain realized.

If an undivided interest in a property is sold, the fraction or percentage of IDC (and depletion for properties placed in service after 1986) allocable to the interest sold is subject to recapture. If part of a property (other than an undivided interest) is sold, all IDC and depletion subject to recapture are applied to the portion sold first. If the gain on a disposition of a part of a property (other than an undivided interest) is less than the total IDC and depletion subject to recapture, the remaining IDC and depletion are subject to recapture on the sale of the balance of the property. If a taxpayer can satisfy the IRS that IDC incurred on a property does not relate to the interest sold, the foregoing rules do not apply.

A question exists as to whether IDC has to be recaptured if a nonoperating interest is sold since a nonoperating interest is not a "property" for purposes of the IDC recapture rules. While it appears clear that depletion must be recaptured on the sale of a nonoperating interest placed in service after 1986, IDC should not be subject to recapture as a result of such a disposition. However, the present status of the rules relating to nonoperating interests is unclear.

Recapture of IDC and depletion may also occur in dispositions other than sales. There are, however, several exceptions to the recapture rules which should be reviewed for these other transactions.

Uniform Capitalization Rules

As discussed in Chapter 5, the general uniform capitalization rules in the Internal Revenue Code require that certain overhead items be capitalized. IDC is specifically exempted from the uniform capitalization rules. Nevertheless, the IRS has concluded that an allocable share of overhead expense must be included as part of IDC. Including overhead as part of IDC increases the amount of IDC which may be subject to the AMT described in Chapter 14.

Turnkey Drilling Contracts

A final problem which must be considered for IDC is allocating costs incurred under a turnkey contract. Under a turnkey drilling contract, the driller is paid a specified amount for providing the drilling rig and crew and for drilling and equipping a well to agreed specifications. A problem may arise in allocating the total contract price between IDC and equipment, unless specific amounts for equipment and IDC are set forth in the contract. A problem arises also if a turnkey contract for exploration or development of a property provides not only for drilling and equipping, but also provides for the acquisition of an interest in the property that is subject to the contract. To avoid allocation disputes with the IRS, the contract should include an allocation of costs between IDC, equipment, and leasehold (if applicable).

Depreciable Equipment

As discussed above, IDC does not include expenditures for tangible property ordinarily considered to have a salvage value. The cost of tangible equipment is recovered through some form of depreciation or cost recovery.

A depreciation deduction is allowed only if the taxpayer possesses a beneficial interest in the equipment (although the taxpayer need not hold legal title) and the taxpayer has an investment, or tax basis, in the depreciable property. The holder of a nonoperating interest, such as a royalty, overriding royalty, net profits interest, or production payment is not entitled to depreciation of equipment because the nonoperating interest owner has no beneficial interest in the equipment.

In addition to lease and well equipment, items such as gathering systems, salt water disposal facilities, nonrecoverable cushion gas, and "line pack" gas which must be present in a pipeline at all times are depreciable assets.

If an oil and gas company purchases a working interest in a producing property, the amount paid must be allocated between leasehold cost and depreciable cost on the basis of relative fair market values.

At the present time, most depreciable property placed in service after 1986 is depreciated under the Modified Accelerated Cost Recovery System (MACRS) which uses statutory methods to recover all of the cost (salvage is disregarded) of property using depreciable lives set forth in the Internal Revenue Code. New and used property are treated in the same manner.

MACRS applies to all depreciable property, other than property the taxpayer elects to expense or amortize, as well as property depreciated under a method not based upon a term of years, such as the unit-of-production method.

As to the election to expense otherwise depreciable costs, such election is available to a taxpayer who does not place in service assets in excess of $200,000 in a taxable year. That taxpayer may currently expense up to

$10,000 of otherwise depreciable cost for the year.

Property placed in service after 1980 and before 1986 is subject to the Accelerated Cost Recovery System (ACRS). Property placed in service prior to 1981, and certain other limited categories of property, are subject to the traditional depreciation rules (e.g., straight line, declining balance, etc.).

Under MACRS, lease and well equipment is included in the seven-year class. The equipment is depreciated using either a 200 or 150 percent declining-balance method, with a change being made to straight line in the taxable year for which it yields a larger deduction. Normally, a half-year convention is used for personal property so that an asset is deemed to have been placed in service, or disposed of, in the middle of the taxable year in which it was placed in service, or disposed of, by the taxpayer.

A taxpayer may elect to use straight line depreciation, rather than the declining balance methods, on one or more classes of assets placed in service during the taxable year. Once the election is made, it applies to all property in the class placed in service during the year and is irrevocable.

MACRS contains a number of other detailed provisions which are beyond the scope of this book, but which should be reviewed carefully for categories of oil and gas assets not included in the lease and well equipment category.

Unit-of-Production Depreciation

As indicated above, MACRS does not apply to taxpayers using the unit-of-production method for depreciation. To use the unit-of-production method for depreciating lease and well equipment, the taxpayer must be able to ascertain the oil and gas reserves applicable to the specific unit of property involved since this method ties the depreciation deduction to the level of oil and gas production for the year.

The unit-of-production method divides the year-end adjusted tax basis of equipment, net of salvage value, by the estimated recoverable reserves at the start of the year. Hence, a per unit depreciable amount is computed. The per unit rate is multiplied by the units of current year production to determine the allowable depreciation deduction for the year.

In determining the amount of reserves used in the calculation of the depreciation deduction, only known oil and gas reserves should be taken into account. When new wells are drilled and reserves increased, the amount of reserves used in the depreciation calculation should be adjusted to reflect the new reserves.

Asset Transfers

From time to time, equipment is transferred from one property to another by a taxpayer. No gain or loss results for Federal income tax purposes from these transfers. Further, no loss is recognized from normal retirements of part of the equipment in a particular group of assets. The cost of the item retired is credited to the asset account and charged or debited (net of salvage) to the depreciation reserve. If equipment is disposed of because of obsolescence or other unusual reasons, a gain or loss may be recognized. In calculating the gain or loss, the tax basis is computed by calculating the depreciation for the equipment as if the equipment had not been depreciated in a group account.

Depreciation Recapture

If depreciable real or personal property is disposed of by a taxpayer, all or part of the gain realized must be treated as ordinary income to the extent of the lesser of (1) prior depreciation subject to recapture or (2) the gain realized.

All depreciation on personal property is subject to recapture, as is depreciation on certain types of real estate. For most real estate, however, the amount of depreciation subject to recapture is limited to the amount of prior depreciation claimed in excess of straight line depreciation.

Recapture of depreciation occurs in dispositions other than sales. However, several exceptions to the recapture rule exist which should be reviewed for these transactions.

CHAPTER 11

Operating Expenses

General

In operating an oil and gas property, ordinary and necessary business expenses are incurred in day-to-day operations. In addition, certain capital expenditures are incurred. For Federal income tax purposes, the normal rules for capitalization and expense are applied. The operator must, with certain exceptions such as IDC, capitalize any amount paid for equipment buildings, permanent improvements or betterments which increase the value of the property.

Uniform
Capitalization Rules

In 1986, Congress enacted general uniform capitalization rules to provide a standard set of rules to determine costs required to be capitalized, or included in inventory cost. With certain exceptions, the uniform capitalization rules apply to all taxpayers who are engaged in (1) manufacturing, constructing, or engaging in other types of activities involving the production of real property or tangible personal property, or (2) acquiring or holding property for resale.

Exploration and development of an oil and gas property appear to be within the meaning of "production of real property" for purposes of these rules. However, a "production" activity logically does not begin until a well is spudded. G&G and other expenses incurred prior to the spud date should be exempt from these rules, but the present position of the IRS on this point, as mentioned in Chapter 5, is unclear. As indicated in Chapter 10, IDC are specifically excluded from the uniform capitalization rules. Other expenses incurred during the exploration and development period, however, may be affected by these rules.

Under the uniform capitalization rules, a taxpayer is required to capitalize not only direct costs, but also an allocable portion of most indirect costs benefitting an asset produced by the taxpayer, or acquired by the taxpayer for resale, including general administrative and overhead costs. One significant item that may adversely affect the oil and gas industry is the apparent requirement that interest incurred on a property during the exploration and development period be capitalized under these rules.

Because of the broad implications of the uniform capitalization rules and the present uncertainty as to how those rules apply to the oil and gas industry, care should be taken to assure that the rules are properly applied in the day-to-day operations of an oil and gas company.

Secondary and Tertiary Recovery Costs

During the producing life of an oil and gas well, secondary or tertiary recovery programs may be instituted to increase or maintain production. Secondary recovery techniques generally rely on the injection of water or gas to maintain reservoir pressure or to encourage migration of oil and gas in certain specific directions. Tertiary recovery methods generally rely on the injection of chemically treated substances to recover additional oil and gas.

As indicated in Chapter 10, the drilling of an input well, even in a secondary or tertiary program, should be treated as incident to and necessary for the production of oil and gas, and the cost of the input well should be deductible as IDC.

The cost of water injected in a secondary recovery project should be currently deductible as an ordinary and necessary business expense, although the uniform capitalization rules may ultimately be interpreted to require that this expenditure be capitalized and amortized over the remaining life of the reservoir.

If, in a secondary recovery operation, a taxpayer purchases hydrocarbons for injection into the reservoir, these hydrocarbons may or may not be recovered through subsequent production. If a taxpayer can show that the injected hydrocarbons will never be recovered, the IRS takes the position that the cost of unrecoverable hydrocarbons must be capitalized and depreciated over the life of the reservoir. In cases where it can be shown that the injected hydrocarbons no longer benefit production, the undepreciated cost of the unrecoverable hydrocarbons is deductible as a loss in the year the determination is made.

A more proper treatment of injected hydrocarbons which are not recoverable would seem to be as ordinary and necessary business expenses in the year paid or incurred. If that position is taken, however, it can be expected that the IRS will challenge the immediate deduction of such expenses. Further, the effect, if any, of the uniform capitalization rules on this type of expense cannot be accurately predicted at this time.

With respect to injected hydrocarbons which can be recovered and sold, the cost of the injected hydrocarbons can be treated as cost of goods sold in the year the injected hydrocarbons are deemed sold. A question which has not been resolved in the courts is whether a taxpayer can currently deduct the excess of the current cost of injected hydrocarbons which are recovered in the future over the estimated present value of the hydrocarbons (using the estimated recovery date or dates for purposes of computing present value). If such a deduction is claimed, litigation might be required to sustain it, and there is no assurance that a court will allow such a deduction.

A taxpayer engaged in tertiary recovery operations may take a current deduction for certain qualified tertiary injection expenditures. Qualified tertiary injection costs include any amounts paid or incurred for any tertiary injectant (other than a hydrocarbon injectant which is recoverable) used as part of a tertiary recovery method. Among the methods treated as tertiary recovery methods are (1) alkaline or caustic flooding, (2) conventional steam drive injection, (3) cyclic steam injection, (4) immiscible gas displacement, (5) in situ combustion, (6) microemulsion flooding, (7) miscible fluid displacement, (8) polymer augmented waterflooding, and (9) unconventional steam drive injection. As with certain other expenses discussed in this chapter, the potential impact of the uniform capitalization rules on tertiary operations is not entirely clear at this time.

Removal of Offshore Platforms

Another expense incurred by some taxpayers in the oil and gas industry is the cost of removing offshore platforms and well fixtures after a lease becomes fully depleted. The IRS takes the position that an accrual basis taxpayer may deduct the costs of removing offshore platforms and well fixtures in the taxable years during which the removal services are performed, regardless of whether the work is done by the taxpayer or by a third party contractor.

C H A P T E R 12

Depletion

General

The removal of oil and gas from the reservoir gradually diminishes the quantity remaining in the reservoir until oil and gas reserves are exhausted. Exhaustion of the reservoir is referred to as "depletion." For Federal income tax purposes, depletion is based on the cost of the units produced and sold or the income derived from production. No depletion is allowed for oil and gas destroyed before sale, although a loss deduction may be permitted for a portion of the basis attributable to the part of the deposit destroyed if the taxpayer can prove a clearly identifiable event and the amount of loss.

Depletion is allowed to the holder of an economic interest (as defined in Chapter 3) in an oil and gas property for which depletion is being computed. If a taxpayer not possessing an economic interest in a property receives a part of the proceeds from the sale of oil and gas from that property, the taxpayer has ordinary income not subject to depletion.

The two methods of computing the depletion allowance provided in the Internal Revenue Code are cost depletion and percentage depletion. Cost depletion allows a deduction for the tax basis of an oil and gas property over the life of the property based on the oil and gas produced and sold. On the other hand, percentage depletion is a statutory provision allowing a noncash deduction for a specified percentage of gross income from the property, not to exceed a specified percentage of the taxable income from the property.

A taxpayer must compute both cost depletion and percentage depletion each year and must deduct the larger of the two amounts. Allowable depletion, being the larger of cost or percentage depletion, reduces the tax basis in the oil

Hard at work on the depletion regs...

and gas property. Allowable depletion is not limited to recovery of tax basis, but cost depletion is zero after the tax basis has been fully recovered. Percentage depletion, if allowable, may continue to be claimed after the tax basis has been fully recovered.

Cost Depletion

Cost depletion for Federal income tax purposes allows the recovery of the tax basis of a property in the ratio that current unit sales of oil and gas bear to the total anticipated unit sales of oil and gas to be made over the life of the property. To compute cost depletion, the tax basis at the end of the taxable year (adjusted for prior years' depletion, but not for the current year's depletion) is determined. Next, the reserves attributable to taxpayer's interest in the property as of the end of the taxable year are determined. Only known reserves should be used in the calculation. If a revision in estimated reserves occurs, the revision is taken into account in the year the revision takes place. Lastly, the number of units of oil and gas sold by the taxpayer in the taxable year is determined. Using all of the foregoing information, a per unit cost depletion amount is calculated and multiplied by the number of units produced during the taxable year to determine the cost depletion deduction.

Cost Depletion on Lease Bonuses

Cost depletion may be calculated on a lease bonus using the same formula as outlined above except that monetary amounts are substituted for physical quantities of oil and gas. Cost depletion is computed by first calculating the ratio of the bonus received to the total of the bonus received, plus estimated future royalties from the property. The resulting percentage is multiplied by the tax basis of the retained royalty, or other nonoperating interest, to compute the cost depletion deduction.

In almost every case involving a lease of an unproven property, it is difficult to compute cost depletion because of the uncertainty of estimated future royalties. In one case, the taxpayer estimated future royalties to be zero and claimed cost depletion equal to 100 percent of its tax basis in the property. The IRS, however, contends that if a taxpayer estimates future royalties to be zero, the taxpayer is explicitly indicating that the property does not have an oil and gas deposit and, hence, no cost depletion is allowable. The IRS's position may be subject to further litigation in the future. If a proven property is subleased, it should be possible for the sublessor to estimate future royalties, and claim cost depletion if the taxpayer has tax basis in the retained nonoperating interest.

The foregoing rules also apply to the calculation of cost depletion on minimum or advance royalties.

Percentage Depletion

Prior to the enactment of legislation in 1975, percentage depletion was generally available to all oil and gas producers. After the 1975 legislation, however, percentage depletion was eliminated for some taxpayers and the income upon which percentage depletion is allowed was substantially limited for others.

As indicated above, percentage depletion is a percentage of gross income from the property, limited to a specified percentage of net income from the property. The determination of gross income from the property was discussed in Chapter 9.

Prior to 1991, percentage depletion was either 15 percent or 22 percent of gross depletable income limited to 50 percent of the taxable income from the property. For 1991 and thereafter, the net income limitation is 100 percent of the taxable income from the property and, in some very limited cases, after 1990, the 15 percent rate may be increased, as discussed below.

Taxable income from the property means gross income from the property minus all allowable deductions (excluding the depletion deduction) attributable to the property including the operating expenses, certain selling expenses, administrative and financial overhead, depreciation, taxes, losses, IDC, and similar deductions. Expenditures attributable to both the oil and gas producing and other activities are apportioned between the oil and gas producing activity and the other activities. In addition, deductions which are not directly attributable to a specific oil and gas property are allocated among all properties in computing taxable income from the property.

Expenses directly related to the property are not ordinarily difficult to identify. IDC is readily ascertainable. However, IDC incurred in deepening a well to a new potential deposit does not need to be taken into account as a direct deduction in computing percentage depletion on a producing deposit in the same tract or parcel of land, if an election has been made to treat each deposit as a separate property as discussed in

Chapter 3. This holds true even if the potential deposit proves nonexistent.

Complications can arise in the allocation of indirect expenses to specific properties. Indirect expenses are allocated to various activities of the taxpayer, and those allocated to the producing activity are then allocated to individual producing properties. Indirect expenses include general overhead expenses, including interest incurred in oil and gas operations, and other general expenses. Non-business expenses are not included as part of general expenses.

The method of allocating overhead is not specified in the Internal Revenue Code, or in the administrative promulgations of the IRS. Commonly, indirect expenses are allocated on the basis of direct expenses. Under this procedure, the total indirect expenses are determined and the total direct expenses are determined on a property-by-property basis. The indirect expenses are first allocated among the various activities of the taxpayer, such as exploration, production, transportation, refining, and marketing. Direct expenses of each activity may be used as a basis for the allocation. Indirect expenses assigned to the production activity may be allocated directly to the properties on the basis of direct expenses incurred on the individual properties. Other methods of allocation may be used provided the method is logical and followed consistently.

In 1975, percentage depletion was repealed for all domestic and foreign oil and gas production with certain limited exceptions. Exceptions which remain available today include domestic "fixed contract gas" and a daily barrel exemption for oil and other gas production for independent producers and royalty owners (the "IPRO" exception).

Fixed Contract Gas

Percentage depletion is available on fixed contract gas at 22 percent on unlimited quantities of such production. Fixed contract gas is domestic gas sold under a contract in effect on February 1, 1975, and all times thereafter, under which the price cannot be increased to any extent to reflect an increase in the producer's Federal income tax liability resulting from the repeal of the percentage depletion. Any price increases after February 1, 1975, are presumed to be for the increase in tax liability due to the repeal of percentage depletion unless the taxpayer can show clear and convincing evidence to the contrary. While the price of fixed contract gas cannot be changed to take into account the repeal of percentage depletion, all changes of price do not automatically disqualify the gas as fixed contract gas. Examples of permitted increases include specifically defined dollar amount adjustments not relating to percentage depletion, such as adjustments to reflect (1) higher state and local property or severance taxes, (2) increased operating costs, (3) the highest area rate fixed by a Federal agency for interstate sales, (4) the quality of gas sold, and (5) normal escalations.

Independent Producers and Royalty Owners (IPRO) Exemption

In addition to the exemption for certain domestic natural gas, IPRO allows percentage depletion on stated volumes of domestic oil and gas production. It is important to note that fixed contract gas does not count in determining the production qualifying for percentage depletion under IPRO.

Under IPRO, the depletable oil quantity is limited to 1,000 barrels of crude oil production per day and the percentage depletion rate is 15 percent. In order to calculate percentage depletion on domestic natural gas (that is, gas other than fixed contract gas) under IPRO, an election must be made to convert some portion or all of the depletable oil quantity into an equivalent amount of depletable gas quantity. The depletable gas quantity of any taxpayer in any tax year is equal to 6,000 cubic feet of natural gas multiplied by the number of barrels of the taxpayer's depletable oil quantity for which an election to convert has been made. The portion of the oil quantity converted reduces the amount of oil upon which the producer can claim percentage depletion. The conversion election is made annually by claiming percentage depletion for the taxable year based upon the election. No separate election statement is required.

In computing a taxpayer's average daily production of domestic crude oil and natural gas, the aggregate production of domestic oil or natural gas during the year is divided by the number of days in the taxable year, thereby averaging large increases or decreases during the year. As previously stated, fixed contract natural gas is not considered in computing average daily production.

If a taxpayer's average daily production exceeds the allowable depletable quantity, a special computation is made. First, percentage depletion must be computed on each property regardless of the 1975 limitation. If cost depletion applies on a particular property, no further action is taken for that property and cost depletion is the deductible amount. The tentative percentage depletion from all properties, except those to which cost depletion applies, is multiplied by a fraction, the numerator of which is the current depletable quantity (1,000 barrels per day) and the denominator of which is the taxpayer's average daily production for the year. At this point, a number of additional calculations may be required.

Where related parties are engaged in the production of oil and gas, the depletable quantities must be allocated among the parties. Businesses under common control, component members of a controlled group of corporations, and certain members of the same family must make such an allocation. Detailed constructive ownership and other rules apply in making the allocation of the depletable quantities.

The IPRO definition is used in the recent AMT changes (see Chapter 14), under which "independent producers" do not treat excess percentage depletion as an alternative minimum tax preference item and may limit this amount of IDC as a perference item.

Refiners and Retailers

Certain taxpayers do not qualify for the IPRO because of their business activities.

Refiners are not allowed to claim any percentage depletion under IPRO.

A refiner was defined as a taxpayer that engages in the refining of crude oil and the daily refinery runs of the taxpayer do not exceed 50,000 barrels *on any day* of the taxable year. The mere fact that a related person engages in refining activities of the required size appears to preclude a taxpayer from claiming percentage depletion under IPRO despite the fact that the taxpayer has no control over the activities of the related party.

The Internal Revenue Code also denies percentage depletion under IPRO to any taxpayer who directly, or through a related person, sells oil or natural gas, or any derivative product, as a retailer. Retail sales by a party obligated to use an oil and gas producer's trade name or trademark constitute retail sales by that producer. In addition, sales of specified products through a retailer occupying a retail outlet owned, or under the control of, a producer constitute retail sales by the producer. Certain complex exceptions to the retailer rule may apply to assist a producer in avoiding being treated as a retailer and losing percentage depletion under IPRO.

Transferred Property

Under the 1975 legislation, percentage depletion was not available to transferees in most cases after a proven property was transferred. For transfers after October 11, 1990, the transfer rule has been repealed. Transfers of oil and gas properties after that date will no longer cause loss of percentage depletion. Further, it appears that certain taxable or non-taxable exchanges occurring after October 11, 1990, may re-qualify properties that were previously precluded from using percentage depletion. The IRS should issue regulations clarifying how properties which have been previously transferred may be re-qualified for percentage depletion in transfers after October 11, 1990.

Marginal Property
Percentage Depletion

Another new provision enacted in 1990 may allow percentage depletion on marginal properties to be raised above 15 percent in certain limited circumstances. Under the new provision, percentage depletion is raised by one percentage point for each whole dollar by which $20 exceeds the reference price for crude oil for the calendar year preceding the calendar year in which the taxable year begins. The maximum percentage depletion rate under this provision is 25 percent (a crude oil reference price of less than $10 per barrel would be required to reach that level).

Properties subject to this provision include stripper wells producing oil and gas. These properties are defined as wells for which the average daily production is 15 barrels (equivalent) or less. Stripper status is determined annually. In addition, properties from which substantially all the production is heavy oil (defined as crude oil with a weighted average gravity of 20 degrees API or less, corrected to 60 degrees Fahrenheit) will also qualify as marginal properties under this provision.

Unless a taxpayer elects otherwise, marginal production is given preference over other production for purposes of the 1,000 barrels per day depletable oil quantity under IPRO limitation. However, the election to allocate production ratably over all properties may be beneficial if use of the higher percentage depletion rate attributable to marginal properties will reduce percentage depletion because of the net income limitation. This provision is effective for taxable years beginning after 1990.

The 65 Percent Limitation

Another limiting provision relating to percentage depletion is that the maximum deduction which may be claimed under IPRO is limited to 65 percent of a taxpayer's taxable income, as adjusted, for the taxable year. The limitation has no impact on the amount of cost depletion which may be claimed, nor does it affect the amount of percentage depletion allowable for fixed contract price gas. Only the percentage depletion allowed under IPRO may be reduced.

If the 65 percent limitation applies, the disallowed amount is carried over to the succeeding taxable year, and, presumably, can be carried over for an indefinite period of time.

In order to calculate the 65 percent limitation, percentage depletion calculations are made as if the 65 percent limitation did not exist. Several rather complicated adjustments are made to determine adjusted taxable income. After that determination is made, 65 percent of adjustable taxable income becomes the amount of percentage depletion that can be claimed under IPRO. If the 65 percent limitation applies, another complicated series of calculations is made, and numerous calculations may be required to calculate the amount of actual percentage depletion allowable for a particular property.

If a taxpayer has a percentage depletion carryforward because of the 65 percent limitation, the carryforward may be used in a future year if the total of regular percentage depletion under IPRO for the year, plus the portion of the carryforward used, does not exceed 65 percent of the adjusted taxable income from the taxable year. While the percentage depletion carryforward is maintained by property, the carryforward does not affect the computation of percentage depletion for properties in the later years.

If a property on which percentage depletion is limited because of the 65 percent rule is subsequently disposed of by a taxpayer, the amount of percentage depletion carryforward must be reduced under a rather complex formula.

Example of
Depletion Calculation

Assume a working interest owner drills and completes a productive well in January 1993. During 1993, the well produces 30,000 barrels of oil which result in gross revenues (net of royalties) of $600,000. Total expenses on the property in 1993 are $520,000, resulting in net income from the property of $80,000. The leasehold cost is $300,000 and total reserves in January 1993 are estimated to be 300,000 barrels. The working interest owner's taxable income for calendar year 1993 (before depletion) is $120,000. The allowable depletion deduction is computed as follows:

Cost Depletion
1993 Production

Leasehold cost ÷ Beginning reserves = $300,000 ÷ 300,000
Units produced 1993 = 30,000
Cost depletion = $ 30,000

Percentage Depletion

Gross revenue	$600,000
x Statutory rate	x .15
Percentage depletion before limitation	$ 90,000
Limited to net income from the property	$ 80,000
Taxable income before depletion	$120,000
x 65% limitation	x .65
Taxable income limitation	$ 78,000
Percentage depletion after limitations	$ 78,000

Allowable Depletion

Allowable depletion (greater of cost or percentage depletion)	$ 78,000

CHAPTER 13

Tax Credits

General

For many years, oil and gas producers were entitled to the investment credit for tangible personal property used in oil and gas operations. The investment credit has now been repealed, but certain other credits remain available.

Nonconventional Fuel Credit

The nonconventional fuel credit (often referred to as the Section 29 credit) is a non-refundable income tax credit in the maximum amount of $3 per "barrel-of-oil equivalent" available for the production and sale of alternative fuels to unrelated persons. Barrel-of-oil equivalent means the amount of the fuel produced that has the energy equivalent (expressed in BTU's) of one barrel of oil.

The credit is available for sales before January 1, 2003, from wells drilled or facilities placed in service between 1980 and January 1, 1993. The spud date of a well will determine whether production from the well will qualify for the credit. A facility that produces gas or synthetic fuels from coal will qualify for the credit as long as the facility is placed in service prior to January 1, 1997.

Types of energy sources which are eligible for the credit include (1) oil produced from shale and tar sands, (2) gas produced from geopressured brine, Devonian shale, coal seams, a tight formation or biomass, (3) liquid, gaseous or solid synthetic fuels produced from coal (including lignite) including such fuels when used as

feedstocks, (4) qualifying processed wood fuels, and (5) steam produced from solid agricultural by-products (excluding timber by-products).

The $3 nonconventional fuel credit is phased out as the reference price (determined annually) of domestic unregulated oil rises from $23.50 to $29.50, except for the credit relating to tight gas, which remains at $3. A number of very technical provisions are included in the legislation creating this credit which are beyond the scope of this book. Nevertheless, careful planning may allow a taxpayer producing qualifying energy resources to achieve a significant economic and tax benefit from this credit.

This credit has been particularly susceptible to political pressures and has expired and been retroactively reinstated several times.

Domestic Enhanced
Oil Recovery Credit

In 1990, a new enhanced oil recovery credit was enacted which is equal to 15 percent of qualified enhanced oil recovery costs including the cost of:

1. tangible equipment,

2. IDC (a special rule applies for integrated oil companies), and

3. the cost of qualified tertiary injectants described in Chapter 11.

An enhanced oil recovery project is a domestic project that involves the application of one or more tertiary recovery methods (as defined in Chapter 11).

Significant expansions of existing projects after December 31, 1990, are also included.

Certification for the project must be obtained from a petroleum engineer in accordance with rules to be promulgated by the IRS. The nine tertiary recovery methods set forth in Chapter 11 are considered to qualify for the credit, as is immiscible non-hydrocarbon gas displacement. The legislative history of the new credit indicates that the IRS may review the use of polymer-augmented waterflooding and indicate situations where such method will, and will not, qualify for the enhanced oil recovery credit.

The enhanced oil recovery credit phases out when the reference price for crude oil for the preceding calendar year exceeds $28 (as adjusted for inflation), and the credit will phase out completely when the reference price exceeds $28 per barrel (as adjusted for inflation), plus $6.

The enhanced oil recovery credit is not allowable against the AMT (see Chapter 14). Credits may be carried back 3 years (but not prior to pre-enactment years) and carried forward 15 years. The dollar amount allowed as a credit will reduce the amount of deductions, or the amount to be added to tax basis, relative to costs that are taken into account in computing the credit.

As with the nonconventional fuel credit, some taxpayers may be able to effectively use the enhanced oil recovery credit to improve the economic results of qualifying projects with careful planning.

CHAPTER 14

Limitations on Deductions and the Alternative Minimum Tax

General

For many years, some oil and gas drilling activities were tax motivated and, correctly or incorrectly, were viewed as "tax shelters." During the past 15 years, tax reform legislation has attempted to retain IDC and percentage depletion as incentives for exploring and developing oil and gas reserves, while limiting the ability of tax shelters to exploit such incentives. Unfortunately, many of the limitations imposed to curb tax shelters have dramatically affected the availability of the tax incentives to those directly engaged in the oil and gas industry.

Nonrecourse Debt

The use of nonrecourse debt (debt upon which there is no personal liability to the obligor) during the 1960s and early 1970s to finance oil and gas transactions has been the subject of substantial litigation. The courts have found in several cases that nonrecourse debt was not true debt for Federal income tax purposes. In a typical case using nonrecourse debt, drilling and other expenses were incurred by partnerships using the accrual method of accounting with nonrecourse debt being issued in payment of these obligations. The courts found that the nonrecourse liabilities were not fixed liabilities, and the deductions created thereby were not properly accruable (see also the discussion of the economic performance requirement for accrual basis taxpayers in Chapter 10).

At-Risk Rule

Since 1976, the Internal Revenue Code has provided that the amount of loss otherwise allowable for an activity cannot exceed the aggregate amount a taxpayer has at risk for such activity at end of the taxable year. Obviously, if a taxpayer incurred expenses with a nonrecourse note, such expenses would not be deductible because the taxpayer would not be at risk for those liabilities.

Under the at-risk rule, a taxpayer generally is at risk for an activity to the extent of (1) cash and the adjusted tax basis of other property contributed to the activity, (2) personal liability indebtedness incurred in connection with the activity, and (3) the net fair market value of assets (other than assets used in the activity) securing debt incurred in connection with the activities.

Funds borrowed for development of an oil and gas property secured solely by the reserves under the property being developed would appear not to be included in the amount a taxpayer has at risk for the property, and limitations might result on the amount of allowable deductions for that activity. In any event, the statutory provisions requiring a taxpayer to be at risk may substantially limit the availability of the deductions incurred with nonrecourse debt and should be reviewed carefully in transactions where nonrecourse debt is involved.

Passive Loss Limitations

In 1986, significant limitations on the use of losses and credits from "passive activities" were enacted. A passive activity is defined as the conduct of any trade or business in which the taxpayer does not materially participate throughout the year. An individual materially participates only if he is involved in the activity on a regular and continuous basis. Since a limited partner may not participate in the business of a limited partnership, a limited partnership interest is a passive interest, and any loss generated from that entity is subject to the passive loss rule.

Losses and credits arising for passive activities may be used only to offset income from other passive activities. Unused passive losses and credits can be carried forward indefinitely to offset future income from passive activity. Suspended losses (but not credits) from an activity are allowed in full on disposition of the entire interest in the activity.

An exception to the passive loss rule is provided for a working interest owner unless the taxpayer holds that interest in a form which limits the taxpayer's liability regarding the activities relating to such interest. If an oil and gas working interest is held in a general partnership, the general partners can deduct losses from the working interest against other types of income. If, however, the working interest is held in limited partnership form, the limited partner's losses are treated as passive losses which can only be offset against passive income. The same result would apparently apply in the case of oil and gas interests held by a limited liability company. Passive losses from working interest activities cannot be offset against income from nonoperating interests since the income from the nonoperating interests is treated as portfolio income, not as passive income.

A number of complicated rules are involved in determining precisely how passive losses from working interest operations are to be taken into account. If an oil and gas taxpayer is drilling a well in an ownership format which limits liability for that activity, careful attention must be given to the passive loss rules to avoid unanticipated economic and tax consequences.

Alternative Minimum Tax

Since 1986, taxpayers have been subject to the revised AMT. Under this concept, a taxpayer calculates a tentative AMT by applying a flat rate (20 percent for corporations and 24 percent for non-corporate taxpayers) to the excess of alternative minimum taxable income over an exemption, reduced by the alternative minimum tax foreign tax credit, if applicable. Up to 90 percent of AMT can be offset by foreign tax credits calculated under special AMT rules. The taxpayer's regular Federal income tax liability for the year (with certain adjustments) is then subtracted from the tentative AMT to obtain the amount of AMT due.

The AMT is based upon the applicable tax rate (see Appendix D) applied to alternative minimum taxable income (AMTI) which is calculated by taking the taxpayer's regular taxable income and making certain adjustments thereto, including depreciation and other adjustments.

After adjusting regular taxable income, certain "tax preferences" are added back to adjusted taxable income. Prior to 1993, the most significant adjustments for the oil and gas industry were the excess of percentage depletion over the taxpayer's tax basis on a property-by-property basis and the amount by which Excess IDC exceeded 65 percent of net income from oil and gas operations. Excess IDC is the amount of IDC deducted for regular tax purposes in excess of an amount which would have been deducted had the IDC been capitalized and deducted ratably over 120 months, beginning with the month in which production from the property begins. The calculation of IDC taken into account for purposes of the foregoing computations is subjected to a complex set of rules which must be carefully analyzed before drilling activities are consummated. Certain other tax preferences are added back to adjusted taxable income to arrive at AMTI.

For years after 1992, taxpayers qualifying for the IPR deducation will no longer treat excess percentage depletion as a tax preference item and *may* eliminate all or a portion of their IDC as a preference item depending upon the taxpayers facts and circumstances.

Most of the adjustments required in determining AMTI are not unique to the oil and gas industry. The most common adjustments encountered are given below.

1. Net operating loss deductions must be recomputed under special AMT rules which have the affect of generally reducing the deduction.

2. Gain or loss on the disposition of depreciable tangible equipment is recomputed for AMT purposes when different methods of depreciation are used for regular tax purposes and AMT purposes.

3. For purposes of the regular tax calculation, tangible assets are normally depreciated using a special accelerated method for cost recovery over relatively short lives. For AMT purposes, depreciation must be recalculated by using a less accelerated method generally over longer asset

lives resulting in a smaller depreciation deduction for AMT purposes.

4. In the case of a corporation with Adjusted Current Earnings (ACE) in excess of taxable income, the AMT requires 75 percent of the excess to be included in the calculation of AMTI. ACE is a complicated concept generally following the pattern of the calculation of states' tax earnings and profits. In the event an ACE adjustment is contemplated, expert advice must be obtained in determining the impact of calculation.

For taxable years before 1993, an energy deduction is provided for AMT purposes. The deduction allows an independent producer a deduction from alternative minimum taxable income for a specified percentage of exploratory IDC, development IDC and percentage depletion attributable to marginal properties previously added back in determining alternative minimum taxable income. Various percentages apply in determining the amount of each item which can be deducted. The procedure for determining the deduction is complicated, and many unanswered questions exist as to the proper interpretation of the provision. Because of its complexity and limited application and duration, the deduction may prove to be of limited benefit to independent producers, although, in some cases, the deduction may prove beneficial. The "energy deduction" is suspended from 1993 through 1997.

A taxpayer is allowed a credit against subsequent years' regular income tax liabilities for the taxpayer's post-1986 AMT to the extent the AMT is attributable to tax preferences or adjustments that involve the deferral of income or the acceleration of deductions. Generally, a credit is not allowed for AMT attributable to preferences or adjustments involving permanently disallowed deductions or permanent exclusions of income from the tax base.

Without question the AMT has, in years past, severely damaged the value of IDC and percentage depletion as tax incentives. Taxpayers engaged in oil and gas exploration and development must carefully review the AMT provisions to be sure that the deductions that are contemplated will actually have the benefit anticipated and will not be substantially reduced because of the AMT.

Dispositions of Oil and Gas Properties

General

As discussed in Chapter 4, the original owner of the mineral rights may dispose of his interest either by sale or lease. In this chapter, the conveyance to a third party of all or a part of an existing operating or nonoperating interest in a property is reviewed.

Sublease

A sublease arises when the working interest owner assigns the working interest to another party for valuable consideration, but retains a continuing nonoperating interest in the production, such as a royalty or net profits interest. Because of the retention of the continuing nonoperating interest, the transaction is classified as a sublease. The same tax rules (see Chapter 4) apply to a sublease as apply to a lease.

If the subleased property is developed, the sublessor must allocate the consideration received between the equipment and the bonus. Since the owner of a nonoperating interest has no ownership in the equipment, the equipment is deemed to have been sold at its fair market value and some or all of the gain on the equipment may be treated as ordinary income because of the recapture of depreciation rule discussed in Chapter 10. If the cash consideration is less than the tax basis of the equipment, the proper handling of the undercovered basis is subject to two interpretations. The IRS takes the position that no loss is recognized in such a transaction and that the unrecovered basis should be transferred to the depletable basis of the retained nonoperating interest. However, the U.S. Supreme Court has effectively stated that depreciable basis cannot be converted into depletable basis. An argument can be made that a loss can be claimed in the year the equipment is transferred to a third party.

Sale

An oil and gas property is deemed to have been sold (1) when the owner of an interest in the property assigns all of that interest or assigns a fractional share identical (except as to size) to the fractional interest retained or (2) if the working interest owner assigns any type of continuing nonoperating interest in the property and retains the working interest. In addition, if the owner of the operating interest assigns the operating interest and retains a non-continuing, nonoperating interest (a production payment), a sale is deemed to have occurred. The gain or loss from the sale may be taxed as ordinary income or loss, or capital gain or loss, depending on the application of the recapture provisions and whether the taxpayer is a dealer in oil and gas properties.

If the taxpayer is a dealer in oil and gas property, the gain or loss will generally be treated as ordinary gain or loss. Hence, a taxpayer making frequent purchases and sales without carrying on the activities of exploring and developing properties may be categorized as a dealer. Careful record-keeping is necessary to avoid categorization as a dealer if multiple oil and gas sale transactions are carried on by a taxpayer.

If the taxpayer is not a dealer, gain and loss on the sale of a property will generally be capital gain or loss subject to the application of the recapture rules, which may cause some or all of the gain to be ordinary income. In some cases, a loss may be an ordinary loss, rather than a capital loss, depending on the taxpayer's overall gain or loss situation for the taxable year.

A taxpayer may choose to assign only a fractional share of the working interest, in which case gain or loss equal to the difference between the sales price and the part of the tax basis allocable to the interest sold is recognized. In other cases, all the working interest in a divided part of the original lease, or all the rights at some depths, but not all depths, may be sold. The basis of the original property is allocated between the interest sold and the interest retained on the basis of relative fair market values, and gain or loss equal to the difference between the sales price and the allocable portion of the tax basis is recognized.

If a taxpayer sells all or part of the working interest in a developed property, the gain or loss from the sale of the equipment and from the sale of the working interest must be computed separately. As indicated in Chapter 10, gain on the sale of equipment may be treated as ordinary income under the recapture rules. As to the gain on the disposition of the working interest, recapture of IDC (and depletion for a property placed in service after 1986) is required.

Sale of Nonoperating Interest

If all or a part of a nonoperating interest is sold, gain or loss equal to the difference between the sales price and the tax basis allocated to the property is recognized. Creation of an overriding royalty or net profits interest out of a working interest is considered a sale, and gain or loss is measured by the difference between the sale proceeds and the tax basis allocated to the interest sold. Since no portion of the working interest has been sold, the IDC recapture rules should not apply to cause part of the gain to be ordinary income. However, for a property placed in service after 1987, the recapture of depletion rule will apply.

Production Payments

As indicated in Chapter 3, a production payment carved out of the working interest is generally treated as a production loan by the working interest owner and by the acquiror of the production payment. All production income continues to belong to the working interest owner who is entitled to depletion on all production and is obligated for all costs. Amounts paid to the owner of the production payment are treated first as interest, and then as payment of principal.

If a carved out production payment is pledged to the exploration and development of the property from which it is created, no income results, but the proceeds of the production payments are treated as a reduction of costs to which the proceeds are applied. The acquiror of such a production payment is deemed to have acquired a depletable interest in the property. As a result, the share of the production applied to liquidate the production payment is excluded from the working interest owner's income and is reported as gross income to the production payment holder.

Some confusion exists whether carved out production payments pledged to certain development activities are to be treated as an economic interest or as a loan. If the proceeds are pledged and used to install equipment during the drilling period and before production from the property, the carved out production payment qualifies as an economic interest in the hands of the production payment owner. If, however, production has commenced before the production payment proceeds are used, the IRS may argue that the proceeds used to acquire production equipment should be treated as a loan.

If a production payment is carved out for exploration and development, the proceeds must be pledged for that use on the specific property or properties burdened by the production payment in order for the interest to be an economic interest in the hands of the production payment owner. If the proceeds are not used on these properties, but are used for exploration and development of other properties, the production payment is treated as a loan.

If a property is transferred and the owner retains a production payment, the production payment is generally treated as a purchase money mortgage. The transferor is deemed to have sold the property on an installment basis, with the total proceeds being the amount of cash received, plus the retained purchase money mortgage. The transferee of the working interest is the owner of the entire working interest, subject to a purchase money mortgage. As a result, the transferee is taxable on the entire production attributable to the working interest even though part of the proceeds is paid to the seller as mortgage payments.

As discussed in Chapter 4, if a production payment is retained in a leasing or subleasing transaction, it is treated as an economic interest rather than as a mortgage loan. Accordingly, income received by the lessor from the production payment is subject to depletion. The lessor is treated as having given the lessee a bonus payable in installments. Payments by the lessee on the production payment are capitalized as leasehold cost.

Variable royalties are sometimes used in oil and gas transactions. The IRS takes the position that the part of the royalty in excess of its lowest possible percentage at any subsequent time may be a production payment. However, if the royalty varies based solely on a change in the rate of production in a period of one year or less, it is not necessary to treat part of the royalty as a production payment.

Abandoned or Worthless Properties

When a taxpayer has incurred costs for an oil and gas property and the property becomes worthless or the taxpayer abandons the property, a loss deduction is allowed in the year the property becomes worthless or is abandoned. If a taxpayer has incurred deferred G&G costs, a deduction for the deferred costs is allowed in the year that the property to which the G&G pertains is determined to be worthless.

A deduction for a loss attributable to a worthless property is normally allowed in the taxable year in which that loss is actually sustained, provided there is no reasonable prospect for recovery by any means. Formal disposal of a property, or some other overt act of abandonment, is not necessary if worthlessness can be established by some other means. Hence, a loss is allowable if it can be shown that no value whatever remains, even though a sale, abandonment, or other irrevocable loss of title is postponed to a later date. While the courts appear generally to require a relinquishment of title (normally through cessation of payment of delay rentals) for a loss to be allowed for an oil and gas property, the more proper approach of analyzing the availability of the loss is to first determine if the taxpayer has ceased to pay delay rentals or otherwise totally relinquished its rights to the property in the taxable year. If so, an abandonment loss should be allowed without further question (assuming the property did not become worthless in an earlier year). In the event a taxpayer has not ceased payment of delay rentals (or otherwise relinquished title) for a prudent and informed business reason (or reasons), the objective facts of the case at the end of the taxable year in which the loss is claimed should be reviewed. If a taxpayer can provide substantial proof as to the reasons and factors upon which the decision to consider a particular property worthless was made, a loss deduction should be allowed in the year worthlessness is proved, regardless of the fact that the title may not be relinquished until a later date. Unless a taxpayer actually relinquishes title, however, the IRS may assert that the loss is not allowable.

Types of Business Entities

General

Oil and gas operations may be conducted in a number of different types of entities. Among the forms available are (1) proprietorship, (2) general or limited partnership (including a joint operating agreement unless the parties have elected election not to be treated as a partnership), (3) an S Corporation, (4) a limited liability company, and (5) a C Corporation or an association taxable as a C Corporation. See Appendix D for a schedule of U.S. rates for various types of entities

Proprietorship

An individual may carry on oil and gas activities as a proprietorship. In that case, the results are reflected directly in the individual's personal tax return, subject to individual income tax rules and tax rates. If a proprietorship acts together with others for some purposes, as is usually the case in oil and gas ventures, it must be determined whether such individual arrangements are treated as a co-ownership, a partnership, or an association taxable as a corporation for Federal income tax purposes

Partnerships

The Internal Revenue Code provides that a partnership exists where a syndicate, group, pool, joint venture, or other unincorporated organization, through or by

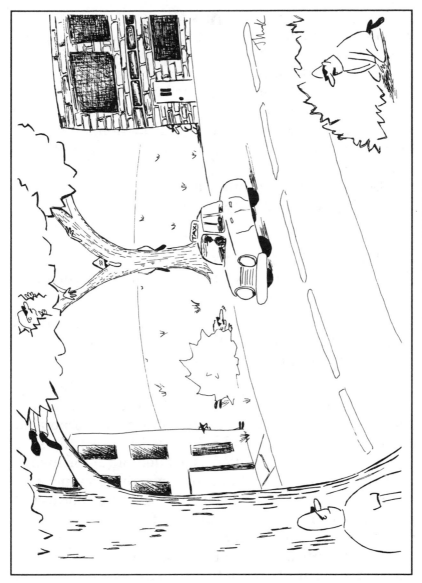

TAXI EVADERS

means of which any business, financial operation, or venture is carried on, and which is not a corporation, trust, or estate. The Federal income tax law definition of partnership is therefore much broader than the ordinary legal meaning of the term for state law purposes.

A partnership is not a taxpayer, but rather serves as a conduit of all items of income, deduction, and credit to its partners who report their allocable share of such items in their own returns.

Generally, for a venture to be treated as a partnership for Federal income tax purposes:

1. the venture must have a profit motive,

2. the profits produced by the venture must be shared jointly by two or more parties involved in the venture, and

3. two or more of the parties sharing the profits must be acting on their own behalf and not as mere agents or employees.

If a joint arrangement constitutes a partnership under the Internal Revenue Code, taxpayers who are participants in the joint venture or other unincorporated organization may elect to be excluded from partnership treatment if the organization is availed of (1) for investment purposes only and not for the active conduct of a business, or (2) for joint production, extraction or use of property, but not for the purpose of selling services or property produced or extracted.

Participants in an oil and gas joint venture normally qualify for the election not to be treated as a partnership so long as they own the property as co-owners, reserve the right to separately take oil and gas in kind or dispose of their shares of the oil and gas produced and do not jointly sell the oil and gas extracted. Each participant is, however, allowed to delegate authority to dispose of its oil and gas to another party, provided (1) the delegation is for no longer than the minimum industry needs require, and in no event for more than one year, and (2) one of the principal purposes of the entity is not for the cycling, manufacturing, or processing of the oil and gas for parties who are not participants.

All participants must agree to make the election not to be treated as a partnership. The election is irrevocable unless the organization ceases to qualify for the election, or unless the IRS consents to revocation of the election. Upon a valid election not to be treated as a partnership, the co-owners must report their respective shares of income and expenses from the venture in their own returns and make their own elections for deducting and capitalizing the various expenses incurred in the venture, including IDC.

Partnership vs. Association

In all oil and gas activities, care must be taken that an arrangement does not create an association taxable as a C Corporation, unless that treatment is desired. The IRS regulations provide six corporate characteristics for determining whether an unincorporated association more nearly resembles a corporation or some other form of entity. The characteristics are:

1. associates,

2. a joint profit objective,

3. continuity of life,

4. centralized management,

5. limited liability, and

6. free transferability of interest.

If an entity has more corporate than non-corporate characteristics, it constitutes an association taxable as a corporation. If the corporate characteristics of an entity are equal to or less than the non-corporate characteristics, the entity will not be treated as an corporation.

Since associates and a joint profit objective are viewed as common to both corporations and partnerships, such characteristics are ignored in determining whether an organization constitutes a partnership or an association taxable as a corporation. Accordingly, if a partnership or joint venture possesses at least three out of the four remaining corporate characteristics, it is categorized as an association taxable as a corporation.

In order to avoid the problems associated with a potential categorization as an association taxable as a corporation, taxpayers may want to consider utilizing a general partnership formed under the Uniform Partnership Act or a limited partnership formed under the Uniform Limited Partnership Act. Entities formed under both of those statutes should be categorized as partnerships for Federal income tax purposes and not as an association taxable as a corporation, provided care is taken in drafting the agreement.

Limited Liability Company

One of the problems with using the general partnership or the limited partnership is that all, or at least one, of the partners must have unlimited liability for partnership activities. During the 1970s, Wyoming created a limited liability company which is similar to a general partnership, except that it provides limited liability for all parties. After a number of years of consideration, the IRS has concluded that such limited liability companies should be treated as partnerships for Federal income tax purposes. Hence, the limited liability company may be a more desirable form of entity than a general partnership or a limited partnership. Caution should be exercised, however, to be sure that the limited liability characteristics is respected in states other than the state of formation for activities carried on outside the state of formation. In addition to Wyoming, several states have enacted or are considering enactment of limited liability company statutes. Because of the growing importance of the limited liability company in the oil and gas industry, additional and more technical information relating to this type of entity is set forth in Appendix E.

S Corporation

Use of an S Corporation to conduct an oil and gas venture allows the parties to obtain the non-tax advantages of the corporate form while avoiding many of the disadvantages of corporate form for Federal income tax purposes. If an S Corporation is used, only one level of Federal income taxation is imposed at the shareholder level. In a like manner, the net losses of an S Corporation flow through and are reported by the individual shareholders.

The primary disadvantage of an S Corporation is its lack of flexibility and organizational structure. To qualify as an S Corporation, certain specific statutory requirements must be met or the organization is treated as a C Corporation for Federal income tax purposes. Normally, to be an S Corporation, a corporation has to be a domestic corporation having no more than 35 shareholders (a husband and wife are treated as one shareholder for this purpose). In addition, the only permissible shareholders are individuals and certain specified estates and trusts. Also an S Corporation cannot have a resident alien as a shareholder and can have only one class of stock, although differences in voting rights alone will not constitute a second class of stock. Lastly, corporations which are members of an affiliated group of corporations are not eligible to make an S Corporation election.

After an S Corporation election has been made, the election may be terminated by several events. The election will terminate if the corporation ceases to satisfy any of the requirements for being an S Corporation. In addition, the election can be revoked if more than 50 percent of the shareholders consent to the revocation. Lastly, if the corporation was a C Corporation before it made its election to become an S Corporation and had accumulated earnings and profits as a C Corporation, the S Corporation election will terminate if the corporation has passive income exceeding 25 percent of the gross receipts for three consecutive years.

If an S Corporation is used, careful attention must be given to the statutory requirements for qualifying and retaining the S Corporation election. In cases where use of an S Corporation is being considered, taxpayers may want to seriously consider use of a limited liability company to avoid the complexities and rigidity of the S Corporation.

C Corporation

In a C Corporation, net income is taxed twice—first at the corporate level and then at the shareholder level when net income is distributed. Net losses accrue to the benefit of the corporation, not its shareholders. Double taxation of net income and the inability to allow the shareholders to claim net losses are two significant disadvantages of operating an oil and gas activity in a C Corporation, although non-tax reasons may dictate the use of a C Corporation in many cases.

CHAPTER 17

International Operations by U.S. Taxpayers

General

A number of U.S. oil and gas companies engage in petroleum exploration and development outside the United States. This chapter will briefly review the general rules applicable to the U.S. companies carrying on such activities.

Foreign Tax Credit

Under the laws of the United States, the entire income of U.S. citizens, domestic corporations and resident aliens is subject to tax in the United States, regardless of source.

Income derived outside the United States may also be subject to taxation in the source country, that is, the country in which the income is earned. Hence, it is possible for U.S. taxpayers to be taxed twice on the same income unless a provision is available to allow a credit against U.S. tax for taxes paid in the source country.

The foreign tax credit mechanism allows U.S. taxpayers to offset allowable foreign taxes paid against the U.S. income tax liability imposed on foreign income. The foreign tax credit, however, may not offset taxes on income earned within the United States. If a taxpayer so elects, foreign taxes may be deducted for U.S. income tax purposes, rather than treated as a credit. However, it is generally more favorable to treat such taxes as a credit. Nevertheless, a U.S. oil company engaging in foreign exploration and development must carefully consider the proper election based upon its own facts and circumstances.

For a tax imposed by a foreign country to qualify as a creditable tax for U.S. income tax purposes, the tax must be a tax on income, profits, or excess profits. Royalties and similar payments must be treated as deductible expenses, rather than as taxes that can be used as U.S. foreign tax credits.

During the course of foreign exploration and development by U.S. oil companies, the determination of whether a particular tax paid to a foreign country is, in fact, an income tax for U.S. foreign tax credit purposes has been one of the most difficult U.S. tax issues for oil and gas companies. The distinction between an income tax on profits and a royalty paid to a foreign government has been the subject of substantial controversy.

In the early years of foreign exploration and development, payments to a number of major oil-producing countries were treated as creditable against U.S. income taxes paid on foreign income. Over the past 20 years, however, the IRS has changed its position, and Congress has enacted rules that significantly limit the availability of foreign tax credits resulting from foreign petroleum operations.

The Internal Revenue Code now provides that foreign income tax which a U.S. oil company pays on foreign oil and gas extraction income qualifies as a creditable tax in the United States only after applying certain restrictive tests. First, a determination must be made whether the payments to a foreign country qualify as creditable taxes for income tax purposes. Then, a determination must be made as to the amount by which otherwise creditable foreign taxes must be reduced by virtue of several very complex provisions in the Internal Revenue Code that apply to foreign petroleum activities. After making that determination, an oil and gas company applies the foreign tax credit limitations set forth in the Internal Revenue Code to determine the amount of foreign tax credit available for a particular taxable year.

If a treaty exists between the potential host country and the United States, a provision may be included in that treaty regarding the definition of a creditable tax. For example, in the Income Tax Treaty between the United Kingdom and the United States, a provision was included regarding the credibility of the United Kingdom Petroleum Revenue Tax in the United States. Another example of a treaty containing such a definition is the treaty between Norway and the United States.

In analyzing the U.S. limitations on the use of foreign tax credits, a taxpayer also should carefully review the rules relating to the recapture of foreign oil and gas extraction losses and other rules relating to the treatment of foreign losses.

The detailed rules and limitations regarding the calculation of the U.S. foreign tax credits and treatment of foreign losses are beyond the scope of this book. However, a U.S. taxpayer contemplating foreign exploration and development activities must be sure that these complicated rules are thoroughly understood and taken into account in choosing the form of entity to be utilized and in developing the economic model for its proposed activities.

Economic Interest

In analyzing how the U.S. income tax rules relating to oil and gas exploration apply to a particular foreign operation, the initial question to be answered is whether the U.S. oil company has an economic interest (see

Chapter 3) in the foreign area, since the ownership of an economic interest is required for income tax payments to be treated as creditable foreign taxes and for certain other U.S. tax provisions to apply.

Normally, under a foreign exploration arrangement, an oil and gas company undertakes the exploration and development of a specific area under a well-defined program. The oil and gas company is obligated to provide funds and equipment for exploration and development and bears the risk of failure. An oil and gas company normally has the right to share in the oil and gas produced or the proceeds from the sale of the production and must look to income derived from production for a return of its capital investment in the exploration and development activity.

The arrangement in most potential host countries is for a fixed period of time. If the foregoing characteristics are present, an oil and gas company should be deemed to have an economic interest in the property for U.S. income tax purposes. Even if the U.S. oil and gas company's rights terminate before the end of the expected economic life of a project, the oil and gas company should, nevertheless, be deemed to own an economic interest provided the oil company is obligated to bear all exploration, development, and production risks and must look solely to production to recoup its investment.

IDC

For IDC incurred after 1986 outside the United States, the Internal Revenue Code provides that such IDC must be recovered (with certain limited exceptions) either over a 10-year period using straight line amortization, or, at the election of taxpayer, as part of the depletable basis in the property. This rule is contrary to the traditional U.S. income tax rule that, upon proper election, IDC can be expensed in the year paid or incurred as discussed in Chapter 10. The capitalization requirement for foreign IDC does not apply to dry hole costs.

In addition to the capitalization rule applicable to IDC, the Internal Revenue Code requires the recapture of IDC as ordinary income on the disposition of a foreign oil and gas property.

Percentage Depletion

As indicated in Chapter 12, percentage depletion was repealed for foreign oil and gas production in 1975. Apparently, Congress decided that percentage depletion should no longer be available to U.S. oil and gas companies for foreign petroleum production. Percentage depletion is available for production on a very limited basis within the U.S. only.

G&G

Oil and gas companies undertaking exploration and development activities in foreign areas must first perform G&G for the potential area of exploration. Obviously, this information is a significant part of the overall evaluation whether or not exploration activities should be undertaken in a particular area. Normally, expenditures for this activity are capitalized for U.S. income tax purposes and become part of the depletable basis of areas acquired, and are expensed for areas that are not acquired, as discussed more fully in Chapter 5. Significant U.S. income tax complications can develop if such expenditures are made in an area and a part of the area is later relinquished because it does not have exploration and development potential.

A question can arise as to what portion, if any, of the G&G attributable to the overall area are allocable to the relinquished area and, therefore, deductible for U.S. income tax purposes. The question occurs with some frequency in foreign areas because many agreements require that certain portions of the original acreage covered by the arrangement be relinquished at specified time intervals. U.S. taxpayers should carefully review the rules applicable to G&G to be sure that appropriate records are kept to maximize deductions that may be realized from future relinquishments.

C H A P T E R 18

Foreign Ownership of U.S. Oil and Gas Operations

General

Normally, foreign investors (whether a corporation or a non-resident alien individual) are not taxed in the United States when assets held in the United States are sold. However, an exception occurs when a foreign investor disposes of an interest in real property in the United States. In such a case, gain realized is treated as income effectively connected to a U.S. trade or business. This gain is taxable in the United States. U.S. real property includes U.S. oil and gas property and tangible equipment associated with the property. An interest in a U.S. corporation is also an interest in U.S. real property if 50 percent or more of a corporation's assets (based upon fair market value) are U.S. real property interests. Interests in U.S. real property held by partnerships are treated as held proportionately by the partners of the partnership.

Tax on Disposition

If a foreign investor is subject to U.S. tax upon the disposition of U.S. real property, the tax is collected through a reporting and withholding procedure. A purchaser of U.S. real property is required to withhold 10 percent of the gross sales proceeds and remit such proceeds to the IRS. The seller must then file a Federal income tax return in the United States reporting the gain or loss realized upon the disposition and pay any additional tax due or request a refund of the withheld amount.

If the seller can demonstrate prior to the sale that little or no gain will result and no U.S. tax will ultimately be due, the IRS may allow the seller a reduction in, or may waive entirely, the 10 percent withholding tax.

A P P E N D I X A

Commentary & Analysis

DOMESTIC ENERGY TAX POLICY— A NEED FOR INNOVATION

by
Frank M. Burke, Jr.*

> *In this article, Burke, who is editor of the Natural Resources Tax Review takes issue with long-standing energy tax policy, the hallmarks of which are the percentage depletion and intangible drilling cost tax expenditures. He points to the gradual erosion of these benefits over the past 60 years as a reason for rethinking how the tax laws can act to encourage domestic exploration. Burke suggests two alternative taxing schemes which, he believes, would encourage capital investment in exploration activities.*

In mid-July, the American Petroleum Institute reported that U.S. oil imports had soared to more than 45 percent of demand, making it likely that the import level

---

*Reprinted, with permission, from 2 Natural Resource Tax Review 243 (1989).

[1] By contrast, in 1983 imported oil represented only 30.5% of total petroleum consumed in the United States. See United States Petroleum Statistics, 1989 Final, Independent Petroleum Assoc. of America (1989).

will surge past 50 percent before the end of 1989.[1] This fact means that the United States has returned to that dangerous level of reliance on foreign petroleum which opens the door to possible significant price increases by foreign producers. President Bush clearly recognized the severity of this problem when signing the Natural Gas Decontrol Act in late July by stating:

> The bottom line is: A strong domestic drilling and producing business is essential for the national security of the United States of America.

The Bush Administration has not taken a strong leadership role in pursuing the solutions to this problem. It is imperative that the Administration immediately begin work with Congress to develop new legislation which will allow much needed capital formation for exploration by the domestic petroleum industry.

Unfortunately, neither the House or the Senate versions of the pending budget deficit reduction bill include provisions addressing the domestic petroleum development issue.[2] The members of Congress who recognize the severity of the domestic petroleum situation seem to have a tenacious desire to continue touting the use of the politically antiquated "percentage depletion" and "intangible drilling costs" tax incentives as the mechanisms to encourage capital formation for the industry.

Percentage depletion and intangible drilling costs have been the principal tax incentives used for the domestic petroleum industry for over 60 years. During that

---

[2] A preliminary version of the Senate Finance Committee's bill did contain a provision, introduced by Senator Bob Dole, that would have provided additional tax benefits for oil and gas production from "marginal properties." This provision was dropped from the bill reported out of the Finance Committee.

period, there has been an ever-increasing perception on the part of the general public that such incentives are "tax giveaways." The continued use of these overworked and misunderstood concepts has done nothing to improve the image of the domestic petroleum industry. Both concepts are out of step with current tax policies and, when mentioned, draw immediate attention and criticism from tax reform activists. As long as percentage depletion and intangible drilling costs remain as the operative incentives in designing potential capital formation legislation for the domestic petroleum industry, *the industry will never achieve the needed level of investment now required to attempt to offset the ever-increasing import problem.*

Because of the escalating reliance on petroleum imports, it is likely that, by the early 1990s, Congress will be seriously questioning the ability of the domestic petroleum industry to provide for the nation's much needed energy security. With much of the undeveloped domestic oil and gas potential lying under Federal lands, some members of Congress will undoubtedly again suggest the need for a Federal oil company to cure the problems which the domestic petroleum industry, at least in their eyes, has not been able to effectively prevent. One only needs to look at the performance of government oil companies in other countries to determine how effective this approach would probably be in the United States.

What is needed now is a domestic petroleum industry that has access to (1) adequate capital to create the necessary level of domestic drilling activity[3] and (2)

---

[3] The number of wildcat wells drilled during 1988 totaled only 3,153 as compared with over 9,100 wildcat wells in 1981. U.S. Petroleum Statistics, note 1, *supra.*

Federal lands having significant oil and gas development potential.

It is absolutely critical that the Administration, Congress and, yes, even the domestic petroleum industry innovate and consider a new system of petroleum taxation that will provide the necessary capital formation impetus. At least two different approaches should receive serious consideration. The first may be summarized as follows:

1. The percentage depletion and the intangible drilling cost reduction would be repealed and replaced by a new system of deductions under which all costs—geological and geophysical, leasehold, intangible drilling and equipment—would be aggregated and amortized over a 60-month period beginning with the month in which drilling commences. All costs incurred after drilling commences would be added to the cost pool and be deducted over the remaining portion of the original 60-month period. In addition, a research and development credit of 20 percent would be allowed for geological and geophysical costs, intangible drilling costs and equipment costs incurred on exploratory wells, and a credit of 10 percent would be allowed for costs incurred to commence or increase secondary and/or tertiary production.

2. A production credit equal to 3 percent of gross income from secondary and tertiary production would be allowed against a producer's Federal income tax liability to encourage continued production for marginal wells.

3. Landowner royalties would be accorded long term capital gain treatment.

4. Since the percentage depletion deduction and intangible drilling cost deduction would be repealed, such items would no longer constitute preference items for minimum tax purposes. The amortization of the cost pool mentioned above would not be treated as a preference item for minimum tax purposes.

5. The passive loss rules would be clarified by expanding the "working interest" exception for oil and gas wells.

A second approach to petroleum tax reform might be as follows:

1. The percentage depletion deduction and the intangible drilling cost deduction would be repealed.

2. Domestic producers would be allowed a deduction from gross income from oil and gas operating interests for amounts "plowed back" into domestic exploration and development in the form of leasehold costs, geological and geophysical costs, intangible drilling costs and equipment costs. That is, if a taxpayer had $1,000,000 of gross income from oil and gas operating interests and plowed back (spent) $400,000 in qualified expenditures, the taxpayer would receive a deduction for $400,000 for Federal income tax purposes. The plow back deduction would be limited to 90 percent of gross income from oil and gas operating interests to be sure that a taxpayer in the oil and gas business could not completely avoid tax by using this deduction. Excess plow back deductions could be carried back and/or forward as appropriate. If considered essential, only expenditures above certain calculated "base period

expenditures" would be allowed as a plow back deduction. Use of the "base period expenditures" concept could encourage increased exploration and development expenditures above those levels incurred in prior years. Also, only a fraction of leasehold costs would qualify as a plow back expenditure to avoid use of excess leasehold inventories as a means of avoiding tax.

3. Landowner royalties would be accorded long term capital gains treatment.

4. Since the percentage depletion deduction and the intangible drilling cost deduction would be repealed, such items would no longer constitute preference items for minimum tax purposes. The deduction for plow back expenditures would not be a tax preference item.

5. The passive loss rules would be clarified by expanding the "working interest" exception for oil and gas wells.

Either new system would provide an attractive capital formation package for the domestic producers. Obviously, much thought must be given to the details for either concept, but the above suggestions should provide the necessary conceptual outline to allow the process of developing a new system of petroleum taxation to commence.

Each of these proposals could be crafted in a manner that satisfies the current revenue neutrality mind-set of Congress. This could be accomplished, for example, by shifting the tax expenditures away from lower-risk production well and allowing more tax incentives for the exploratory or wildcat well.

A new system of petroleum taxation must be broad,

affecting every oil and gas producer in the United States. One of the major problems of changing the existing incentives is that many independent oil and gas producers consider both percentage depletion and the intangible drilling cost deduction to be sacrosanct—an integral part of the economics of their business, although the real economic value of each item has been reduced significantly by tax reform over the last 20 years. Producers refuse to recognize the significant reduction in value of the historical tax incentives and simply are not willing to accept change. Unfortunately, they have influenced the politicians representing them in Washington that change would be disastrous. Hence, oil state politicians continually use percentage depletion and intangible drilling cost deduction as the focal point for attempts to aid the industry with capital formation. Both the producers and the politicians who represent them should understand that the domestic petroleum industry is a mature industry that is burdened by a tax system that needs a good PR man. Hence, it is time to become realistic and consider a new tax system that is in step with current tax policies and which provides tax incentives which are equally as good, or perhaps better than, the existing ones.

The industry, of course, is quite reluctant to enter into a bargaining session with Congress in which it must give up its long time incentives without a clear signal as to what, if anything, may replace them. Congress, on the other hand, is wary of "tax giveaways" and will scrutinize any new proposal carefully in this era of budget crises. Because of the controversial nature of the subject and the reluctance of the parties to deal with it effectively, the proper approach to solving the problem may be to utilize the same approach as was used for Social Security. That is, appoint an independent panel made up of knowledgeable people who study the matter and recommend the proper course of action. This approach diffuses a politically charged subject and facilitates a reasonably quick recommendation as to how to deal with a difficult and important national problem. In any event, regardless of the approach taken to attacking the problem, the time has come for all parties to attempt to deal with it before we are again faced with an energy crisis.

A P P E N D I X **B**

Producers 88-Revised — 8-42
Texas Standard Form

No._____

Oil, Gas and Mineral Lease

From

_____

_____

To

_____

_____

Dated _____, 19 _____
No. Acres _____
_____County, Texas
Term _____

This instrument was filed for record on the _____ day
of _____ 19__, at _____ o'clock _____ M.,
and duly Recorded in Book _____, Page
_____ of the _____ records of this office.

_____
County Clerk,

_____County, Texas
By _____, Deputy

When recorded return to

The Odee Company, Publishers, Dallas

Producers 88 Revised 8-42 — Texas   Texas Standard Form

Oil, Gas and Mineral Lease

THIS AGREEMENT made this _____ day of _____ 19__, between _____

_____

Lessor (whether one or more), and

_____

Lessee, WITNESSETH:

1. Lessor in consideration of _____ DOLLARS ($_____) in hand paid, of the royalties herein provided, and of the agreements of Lessee herein contained, hereby grants, leases and lets exclusively unto Lessee for the purpose of investigating, exploring, prospecting, drilling and mining for and producing oil, gas and all other minerals, laying pipe lines, building tanks, power stations, telephone lines and other structures thereon to produce, save, take care of, treat, transport, and own said products, and housing its employees, the following described land in _____ County, Texas, to-wit:

and containing _____ acres, more or less. In the event a resurvey of said lands shall reveal the existence of excess and/or vacant lands lying adjacent to the lands above described and the lessor, his heirs, or assigns, shall, by virtue of his ownership of the lands above described, have preference right to acquire said excess and/or vacant lands, then in that event this lease shall cover and include all such excess and/or vacant lands which the lessor, his heirs, or assigns, shall have

the preference right to acquire by virtue of his ownership of the lands above described as and when acquired
by the lessor; and the lessee shall pay the lessor for such
excess and/or vacant lands at the same rate per acre as
the cash consideration paid for the acreage hereinabove
mentioned.

2. Subject to the other provisions herein contained,
this lease shall be for a term of ten years from this date
(called "primary term") and as long thereafter as oil, gas
or other mineral is produced from said land hereunder.

3. The Royalties to be paid Lessor are: (a) on oil,
one-eighth of that produced and saved from said land,
the same to be delivered at the wells or to the credit of
Lessor into the pipe line to which the wells may be
connected; Lessee may from time to time purchase any
royaty oil in its possession, paying the market price
therefor prevailing for the field where produced on the
date of purchase; (b) on gas, including casinghead gas or
other gaseous substance, produced from said land and
sold or used off the premises or in the manufacture of
gasoline or other product therefrom, the market value at
the well of one-eighth of the gas so sold or used, provided that on gas sold at the wells the royalty shall be
one-eighth of the amount realized from such sale; where
gas from a well producing gas only is not sold or used,
Lessee may pay as royalty $50.00 per well per year, and
upon such payment it will be considered that gas is
being produced within the meaning of Paragraph 2
hereof; and (c) all other minerals mined and marketed,
one-tenth either in kind or value at the well or mine, at
Lessee's election, except that on sulphur the royalty
shall be fifty cents (50¢) per long ton. Lessee shall have
free use of oil, gas, coal, wood and water from said land,

except water from Lessor's wells, for all operations here-under, and the royalty on oil, gas and coal shall be computed after deducting any so used. Lessor shall have the privilege at his risk and expense of using gas from any gas well on said land for stoves and inside lights in the principal dwelling thereon out of any surplus gas not needed for operations hereunder.

4. If operations for drilling are not commenced on said land on or before one year from this date the lease shall then terminate as to both parties unless on or before such anniversary date Lessee shall pay or tender to Lessor or to the credit of Lessor in _____ Bank at _____ (which bank and its successors are Lessor's agent and shall continue as the depository for all rentals payable hereunder regardless of changes in ownership of said land or the rentals) the sum of _____ Dollars ($_____), (herein called rental), which shall cover the privilege of deferring commencement of drilling operations for a period of twelve (12) months. In like manner and upon like payments or tenders annually the commencement of drilling operations may be further deferred for successive periods of twelve (12) months each during the primary term. The payment or tender of rental may be made by the check or draft of Lessee mailed or delivered to said bank on or before such date of payment. If such bank (or any successor bank) should fail, liquidate or be succeeded by another bank, or for any reason fail or refuse to accept rental, Lessee shall not be held in default for failure to make such payment or tender of rental until thirty (30) days after Lessor shall deliver to Lessee a proper recordable instrument, naming another bank as agent to receive such payments or tenders. The down cash payment is consideration for this lease ac-

cording to its terms and shall not be allocated as mere rental for a period. Lessee may at any time execute and deliver to Lessor or to the depository above named or place of record a release or releases covering any portion or portions of the above described premises and thereby surrender this lease as to such portion or portions and be relieved of all obligations as to the acreage surrendered, and thereafter the rentals payable hereunder shall be reduced in the proportion that the acreage covered hereby is reduced by said release or releases. In this connection the above described premises shall be treated as comprising _____ acres, whether there be more or less.

5. If prior to discovery of oil or gas on said land Lessee should drill a dry hole or holes thereon, or if after discovery of oil or gas the production thereof should cease from any cause, this lease shall not terminate if Lessee commences additional drilling or re-working operations within sixty (60) days thereafter or (if it be within the primary term) commences or resumes the payment or tender of rentals on or before the rental paying date next ensuing after the expiration of three months from date of completion of dry hole or cessation of production. If at the expiration of the primary term oil, gas or other mineral is not being produced on said land but Lessee is then engaged in drilling or re-working operations thereon, the lease shall remain in force so long as operations are prosecuted with no cessation of more than thirty (30) consecutive days, and if they result in the production of oil, gas or other minerals so long thereafter as oil, gas or other mineral is produced from said land. In the event a well or wells producing oil or gas in paying quantities should be brought in on adjacent land and within one hundred

fifty (150) feet of and draining the leased premises, Lessee agrees to drill such offset wells as a reasonably prudent operator would drill under the same or similar circumstances.

6. Lessee shall have the right at any time during or after the expiration of this lease to remove all property and fixtures placed by Lessee on said land, including the right to draw and remove all casing. When required by Lessor, Lessee will bury all pipe lines below ordinary plow depth, and no well shall be drilled within two hundred (200) feet of any residence or barn now on said land without Lessor's consent.

7. The rights of either party hereunder may be assigned in whole or in part and the provisions hereof shall extend to the heirs, successors and assigns, but no change or divisions in ownership of the land, rentals, or royalties, however accomplished, shall operate to enlarge the obligations or diminish the rights of Lessee. No sale or assignment by Lessor shall be binding on Lessee until Lessee shall be furnished with a certified copy of recorded instrument evidencing same. In event of assignment of this lease as to a segregated portion of said land, the rentals payable hereunder shall be apportionable as between the several leasehold owners ratably according to the surface area of each, and default in rental payment by one shall not affect the rights of other leasehold owners hereunder. If six or more parties become entitled to royalty hereunder, Lessee may withhold payment thereof unless and until furnished with a recordable instrument executed by all such parties designating an agent to receive payment for all.

8. The breach by Lessee of any obligation arising hereunder shall not work a forfeiture or termination of

this lease nor cause a termination or reversion of the estate created hereby nor be grounds for cancellation hereof in whole or in part save as herein expressly provided. If the obligation for reasonable development should require the drilling of a well or wells, Lessee shall have ninety days after ultimate judicial ascertainment of the existence of such obligation within which to begin the drilling of a well, and the only penalty for failure to do so shall be the termination of this lease save as to ten (10) acres for each well being worked on and/or being drilled and/or producing oil or gas to be selected by Lessee so that each 10-acre tract will embrace one such well.

9. Lessor hereby warrants and agrees to defend the title to said land and agrees that Lessee at its option may discharge any tax, mortgage or other lien upon said land and in event Lessee does so, it shall be subrogated to such lien with the right to enforce same and apply rentals and royalties accruing hereunder toward satisfying same. Without impairment of Lessee's rights under the warranty in event of failure of title, it is agreed that if Lessor owns an interest in said land less than the entire fee simple estate, then the royalties and rentals to be paid Lessor shall be reduced proportionately.

10. If any operation permitted or required hereunder, or the performance by Lessee of any covenant, agreement or requirement hereof is delayed or interrupted directly or indirectly by any past or future acts, orders, regulations or requirements of the Government of the United States or of any state or other governmental body, or any agency, officer, representative or authority of any of them, or because of delay or inability to get materials, labor, equipment or supplies, or on ac-

count of any other similar or dissimilar cause beyond the control of Lessee, the period of such delay or interruptions shall not be counted against the Lessee, and the primary term of this lease shall automatically be extended after the expiration of the primary term set forth in Section 2 above, so long as the cause or causes for such delays or interruptions continue and for a period of six (6) months thereafter; and such extended term shall constitute and shall be considered for the purposes of this lease as a part of the primary term hereof. The provisions of Section 4 hereof, relating to the payment of delay rentals shall in all things be applicable to the primary term as extended hereby just as if such extended term were a part of the original primary term fixed in Section 2 hereof. The Lessee shall not be liable to Lessor in damages for failure to perform any operation permitted or required hereunder or to comply with any covenant, agreement or requirement hereof during the time Lessee is relieved from the obligations to comply with such covenants, agreements of requirements.

IN WITNESS WHEREOF, this instrument is executed on the date first above written.

WITNESSES: _____

_____   _____

_____   _____

Single Acknowledgment

THE STATE OF TEXAS,⎫
COUNTY OF ⎭

BEFORE ME, the undersigned, a Notary Public in and for said County and State, on this day personally appeared known to me to be the person whose name subscribed to the foregoing instrument, and acknowledged to me that he executed the same for the purposes and consideration therein expressed.

Given under my hand and seal of office, this the day of A. D. 19

(L. S.)

_____

Notary Public in and for County, Texas

Single Acknowledgment

THE STATE OF TEXAS,⎫
COUNTY OF ⎭

BEFORE ME, the undersigned, a Notary Public in and for said County and State, on this day personally appeared known to me to be the person whose name subscribed to the foregoing instrument, and acknowledged to me that he executed the same for the purposes and consideration therein expressed.

Given under my hand and seal of office, this the day of A. D. 19

(L. S.)

_____

Notary Public in and for County, Texas

Corporation Acknowledgment

THE STATE OF TEXAS, }
COUNTY OF

BEFORE ME, the undersigned, a Notary Public in and for said County and State, on this day personally appeared   , known to me to be the person and officer whose name is subscribed to the foregoing instrument and acknowledged to me that the same was the act of the said corporation, and that he executed the same as the act of such corporation for the purposes and consideration therein expressed, and in the capacity therein stated.

Given under my hand and seal of office, this the   day of   A. D. 19

(L. S.)

_____

Notary Public in and for   County, Texas

SINGLE ACKNOWLEDGMENT

THE STATE OF TEXAS,

COUNTY OF

BEFORE ME, the undersigned, a Notary Public in and for said County and State, on this day personally appeared

known to me to be the person   whose name   subscribed to the foregoing instrument, and acknowledged to me that   he   executed the same for the purposes and consideration therein expressed.

GIVEN UNDER MY HAND AND SEAL OF OFFICE,

this the   day of   A. D. 19

(L. S.)

Notary Public in and for   County. Texas

SINGLE ACKNOWLEDGMENT

THE STATE OF TEXAS,

COUNTY OF

BEFORE ME, the undersigned, a Notary Public in and for said County and State, on this day personally appeared

known to me to be the person   whose name   subscribed to the foregoing instrument, and acknowledged to me that   he   executed the same for the purposes and consideration therein expressed.

GIVEN UNDER MY HAND AND SEAL OF OFFICE,

this the   day of   A. D. 19

(L. S.)

Notary Public in and for   County. Texas

CORPORATION ACKNOWLEDGMENT

THE STATE OF TEXAS,

COUNTY OF

BEFORE ME, the undersigned, a Notary Public in and for said County and State, on this day personally appeared

, known to me to be the person and officer whose name is subscribed to the foregoing instrument and acknowledged to me that the same was the act of the said

a corporation, and that he executed the same as the act of such corporation for the purposes and consideration therein expressed, and in the capacity therein stated.

GIVEN UNDER MY HAND AND SEAL OF OFFICE,

this the   day of   A. D. 19

(L. S.)

Notary Public in and for   County, Texas

PRODUCERS 88-REVISED—8-42
TEXAS STANDARD FORM

No.

Oil, Gas and Mineral Lease

FROM

TO

Dated   19

No. Acres   County, Texas

Term

This instrument was filed for record on the   day of   19   at   o'clock   M, and duly Recorded in Book   Page   records of this office.

of the   County Clerk, County, Texas

By   Deputy

When recorded return to

The Oske Company, Publishers, Dallas

Back of Oil, Gas, and Mineral Lease

A P P E N D I X C

A.A.P.L. FORM 610 - 1989

MODEL FORM OPERATING AGREEMENT

OPERATING AGREEMENT

DATED

———— , 19 ——— ,

OPERATOR _____

CONTRACT AREA _____

_____

_____

_____

COUNTY OR PARISH OF _____ , STATE OF _____

Courtesy of American Association of Petroleum Landmen

A.A.P.L. FORM 610 - MODEL FORM OPERATING AGREEMENT - 1989

TABLE OF CONTENTS

i

A.A.P.L. FORM 610 - MODEL FORM OPERATING AGREEMENT - 1989

TABLE OF CONTENTS

A.A.P.L. FORM 610 - MODEL FORM OPERATING AGREEMENT - 1989

OPERATING AGREEMENT

1
2 THIS AGREEMENT, entered into by and between _____,
3 hereinafter designated and referred to as "Operator," and the signatory party or parties other than Operator, sometimes
4 hereinafter referred to individually as "Non-Operator," and collectively as "Non-Operators."

WITNESSETH:

6 WHEREAS, the parties to this agreement are owners of Oil and Gas Leases and/or Oil and Gas Interests in the land
7 identified in Exhibit "A," and the parties hereto have reached an agreement to explore and develop these Leases and/or Oil
8 and Gas Interests for the production of Oil and Gas to the extent and as hereinafter provided,

9 NOW, THEREFORE, it is agreed as follows:

ARTICLE I.
DEFINITIONS

12 As used in this agreement, the following words and terms shall have the meanings here ascribed to them:

13 A. The term "AFE" shall mean an Authority for Expenditure prepared by a party to this agreement for the purpose of
14 estimating the costs to be incurred in conducting an operation hereunder.

15 B. The term "Completion" or "Complete" shall mean a single operation intended to complete a well as a producer of Oil
16 and Gas in one or more Zones, including, but not limited to, the setting of production casing, perforating, well stimulation
17 and production testing conducted in such operation.

18 C. The term "Contract Area" shall mean all of the lands, Oil and Gas Leases and/or Oil and Gas Interests intended to be
19 developed and operated for Oil and Gas purposes under this agreement. Such lands, Oil and Gas Leases and Oil and Gas
20 Interests are described in Exhibit "A."

21 D. The term "Deepen" shall mean a single operation whereby a well is drilled to an objective Zone below the deepest
22 Zone in which the well was previously drilled, or below the Deepest Zone proposed in the associated AFE, whichever is the
23 lesser.

24 E. The terms "Drilling Party" and "Consenting Party" shall mean a party who agrees to join in and pay its share of the
25 cost of any operation conducted under the provisions of this agreement.

26 F. The term "Drilling Unit" shall mean the area fixed for the drilling of one well by order or rule of any state or federal
27 body having authority. If a Drilling Unit is not fixed by any such rule or order, a Drilling Unit shall be the drilling unit as
28 established by the pattern of drilling in the Contract Area unless fixed by express agreement of the Drilling Parties.

29 G. The term "Drillsite" shall mean the Oil and Gas Lease or Oil and Gas Interest on which a proposed well is to be
30 located.

31 H. The term "Initial Well" shall mean the well required to be drilled by the parties hereto as provided in Article VI.A.

32 I. The term "Non-Consent Well" shall mean a well in which less than all parties have conducted an operation as
33 provided in Article VI.B.2.

34 J. The terms "Non-Drilling Party" and "Non-Consenting Party" shall mean a party who elects not to participate in a
35 proposed operation.

36 K. The term "Oil and Gas" shall mean oil, gas, casinghead gas, gas condensate, and/or all other liquid or gaseous
37 hydrocarbons and other marketable substances produced therewith, unless an intent to limit the inclusiveness of this term is
38 specifically stated.

39   L. The term "Oil and Gas Interests" or "Interests" shall mean unleased fee and mineral interests in Oil and Gas in tracts

40   of land lying within the Contract Area which are owned by parties to this agreement.

41   M. The terms "Oil and Gas Lease," "Lease" and "Leasehold" shall mean the oil and gas leases or interests therein

42   covering tracts of land lying within the Contract Area which are owned by the parties to this agreement.

43   N. The term "Plug Back" shall mean a single operation whereby a deeper Zone is abandoned in order to attempt a

44   Completion in a shallower Zone.

45   O. The term "Recompletion" or "Recomplete" shall mean an operation whereby a Completion in one Zone is abandoned

46   in order to attempt a Completion in a different Zone within the existing wellbore.

47   P. The term "Rework" shall mean an operation conducted in the wellbore of a well after it is Completed to secure,

48   restore, or improve production in a Zone which is currently open to production in the wellbore. Such operations include, but

49   are not limited to, well stimulation operations but exclude any routine repair or maintenance work or drilling, Sidetracking,

50   Deepening, Completing, Recompleting, or Plugging Back of a well.

51   Q. The term "Sidetrack" shall mean the directional control and intentional deviation of a well from vertical so as to

52   change the bottom hole location unless done to straighten the hole or to drill around junk in the hole or to overcome other

53   mechanical difficulties.

54   R. The term "Zone" shall mean a stratum of earth containing or thought to contain a common accumulation of Oil and

55   Gas separately producible from any other common accumulation of Oil and Gas.

56   Unless the context otherwise clearly indicates, words used in the singular include the plural, the word "person" includes

57   natural and artificial persons, the plural includes the singular, and any gender includes the masculine, feminine, and neuter.

ARTICLE II.

EXHIBITS

58

59   The following exhibits, as indicated below and attached hereto, are incorporated in and made a part hereof:

60   _____ A. Exhibit "A," shall include the following information:

61   (1) Description of lands subject to this agreement,

62   (2) Restrictions, if any, as to depths, formations, or substances,

63   (3) Parties to agreement with addresses and telephone numbers for notice purposes,

64   (4) Percentages or fractional interests of parties to this agreement,

65   (5) Oil and Gas Leases and/or Oil and Gas Interests subject to this agreement,

66   (6) Burdens on production.

67   _____ B. Exhibit "B," Form of Lease.

68   _____ C. Exhibit "C," Accounting Procedure.

69   _____ D. Exhibit "D," Insurance.

70   _____ E. Exhibit "E," Gas Balancing Agreement.

71   _____ F. Exhibit "F," Non-Discrimination and Certification of Non-Segregated Facilities.

72   _____ G. Exhibit "G," Tax Partnership.

73   _____ H. Other:

74

- 1 -

A.A.P.L. FORM 610 - MODEL FORM OPERATING AGREEMENT - 1989

1 If any provision of any exhibit, except Exhibits "E," "F" and "G," is inconsistent with any provision contained in
2 the body of this agreement, the provisions in the body of this agreement shall prevail.

ARTICLE III.
INTERESTS OF PARTIES

A. Oil and Gas Interests:

6 If any party owns an Oil and Gas Interest in the Contract Area, that Interest shall be treated for all purposes of this
7 agreement and during the term hereof as if it were covered by the form of Oil and Gas Lease attached hereto as Exhibit "B,"
8 and the owner thereof shall be deemed to own both royalty interest in such lease and the interest of the lessee thereunder.

B. Interests of Parties in Costs and Production:

10 Unless changed by other provisions, all costs and liabilities incurred in operations under this agreement shall be borne
11 and paid, and all equipment and materials acquired in operations on the Contract Area shall be owned, by the parties as their
12 interests are set forth in Exhibit "A." In the same manner, the parties shall also own all production of Oil and Gas from the
13 Contract Area subject, however, to the payment of royalties and other burdens on production as described hereafter.
14 Regardless of which party has contributed any Oil and Gas Lease or Oil and Gas Interest on which royalty or other
15 burdens may be payable and except as otherwise expressly provided in this agreement, each party shall pay or deliver, or
16 cause to be paid or delivered, all burdens on its share of the production from the Contract Area up to, but not in excess of,
17 _____ and shall indemnify, defend and hold the other parties free from any liability therefor.
18 Except as otherwise expressly provided in this agreement, if any party has contributed hereto any Lease or Interest which is
19 burdened with any royalty, overriding royalty, production payment or other burden on production in excess of the amounts
20 stipulated above, such party so burdened shall assume and alone bear all such excess obligations and shall indemnify, defend
21 and hold the other parties hereto harmless from any and all claims attributable to such excess burden. However, so long as
22 the Drilling Unit for the productive Zone(s) is identical with the Contract Area, each party shall pay or deliver, or cause to
23 be paid or delivered, all burdens on production from the Contract Area due under the terms of the Oil and Gas Lease(s)
24 which such party has contributed to this agreement, and shall indemnify, defend and hold the other parties free from any
25 liability therefor.
26 No party shall ever be responsible, on a price basis higher than the price received by such party, to any other party's
27 lessor or royalty owner, and if such other party's lessor or royalty owner should demand and receive settlement on a higher
28 price basis, the party contributing the affected Lease shall bear the additional royalty burden attributable to such higher price.
29 Nothing contained in this Article III.B. shall be deemed an assignment or cross-assignment of interests covered hereby,
30 and in the event two or more parties contribute to this agreement jointly owned Leases, the parties' undivided interests in
31 said Leaseholds shall be deemed separate leasehold interests for the purposes of this agreement.

C. Subsequently Created Interests:

33 If any party has contributed hereto a Lease or Interest that is burdened with an assignment of production given as security
34 for the payment of money, or if, after the date of this agreement, any party creates an overriding royalty, production
35 payment, net profits interest, assignment of production or other burden payable out of production attributable to its working
36 interest hereunder, such burden shall be deemed a "Subsequently Created Interest." Further, if any party has contributed
37 hereto a Lease or Interest burdened with an overriding royalty, production payment, net profits interest, or other burden
38 payable out of production created prior to the date of this agreement, and such burden is not shown on Exhibit "A," such
39 burden also shall be deemed a Subsequently Created Interest to the extent such burden causes the burdens on such party's
40 Lease or Interest to exceed the amount stipulated in Article III.B. above.

41   The party whose interest is burdened with the Subsequently Created Interest (the "Burdened Party") shall assume and
42 alone bear, pay and discharge the Subsequently Created Interest and shall indemnify, defend and hold harmless the other
43 parties from and against any liability therefor. Further, if the Burdened Party fails to pay, when due, its share of expenses
44 chargeable hereunder, all provisions of Article VII.B. shall be enforceable against the Subsequently Created Interest in the
45 same manner as they are enforceable against the working interest of the Burdened Party. If the Burdened Party is required
46 under this agreement to assign or relinquish to any other party, or parties, all or a portion of its working interest and/or the
47 production attributable thereto, said other party, or parties, shall receive said assignment and/or production free and clear of
48 said Subsequently Created Interest, and the Burdened Party shall indemnify, defend and hold harmless said other party, or
49 parties, from any and all claims and demands for payment asserted by owners of the Subsequently Created Interest.

ARTICLE IV.
TITLES

A. Title Examination:

53   Title examination shall be made on the Drillsite of any proposed well prior to commencement of drilling operations and,
54 if a majority in interest of the Drilling Parties so request or Operator so elects, title examination shall be made on the entire
55 Drilling Unit, or maximum anticipated Drilling Unit, of the well. The opinion will include the ownership of the working
56 interest, minerals, royalty, overriding royalty and production payments under the applicable Leases. Each party contributing
57 Leases and/or Oil and Gas Interests to be included in the Drillsite or Drilling Unit, if appropriate, shall furnish to Operator
58 all abstracts (including federal lease status reports), title opinions, title papers and curative material in its possession free of
59 charge. All such information not in the possession of or made available to Operator by the parties, but necessary for the
60 examination of the title, shall be obtained by Operator. Operator shall cause title to be examined by attorneys on its staff or
61 by outside attorneys. Copies of all title opinions shall be furnished to each Drilling Party. Costs incurred by Operator in
62 procuring abstracts, fees paid outside attorneys for title examination (including preliminary, supplemental, shut-in royalty
63 opinions and division order title opinions) and other direct charges as provided in Exhibit "C" shall be borne by the Drilling
64 Parties in the proportion that the interest of each Drilling Party bears to the total interest of all Drilling Parties as such
65 interests appear in Exhibit "A." Operator shall make no charge for services rendered by its staff attorneys or other personnel
66 in the performance of the above functions.

67   Each party shall be responsible for securing curative matter and pooling amendments or agreements required in
68 connection with Leases or Oil and Gas Interests contributed by such party. Operator shall be responsible for the preparation
69 and recording of pooling designations or declarations and communitization agreements as well as the conduct of hearings
70 before governmental agencies for the securing of spacing or pooling orders or any other orders necessary or appropriate to
71 the conduct of operations hereunder. This shall not prevent any party from appearing on its own behalf at such hearings.
72 Costs incurred by Operator, including fees paid to outside attorneys, which are associated with hearings before governmental
73 agencies, and which costs are necessary and proper for the activities contemplated under this agreement, shall be direct
74 charges to the joint account and shall not be covered by the administrative overhead charges as provided in Exhibit "C."

- 2 -

A.A.P.L. FORM 610 - MODEL FORM OPERATING AGREEMENT - 1989

1   Operator shall make no charge for services rendered by its staff attorneys or other personnel in the performance of the above
2   functions.

3   No well shall be drilled on the Contract Area until after (1) the title to the Drillsite or Drilling Unit, if appropriate, has
4   been examined as above provided, and (2) the title has been approved by the examining attorney or title has been accepted by
5   all of the Drilling Parties in such well.

6   **B. Loss or Failure of Title:**

7   1. Failure of Title: Should any Oil and Gas Interest or Oil and Gas Lease be lost through failure of title, which results in a
8   reduction of interest from that shown on Exhibit "A," the party credited with contributing the affected Lease or Interest
9   (including, if applicable, a successor in interest to such party) shall have ninety (90) days from final determination of title
10   failure to acquire a new lease or other instrument curing the entirety of the title failure, which acquisition will not be subject
11   to Article VII.B, and failing to do so, this agreement, nevertheless, shall continue in force as to all remaining Oil and Gas
12   Leases and Interests; and,

13   (a) The party credited with contributing the Oil and Gas Lease or Interest affected by the title failure (including, if
14   applicable, a successor in interest to such party) shall bear alone the entire loss and it shall not be entitled to recover from
15   Operator or the other parties any development or operating costs which it may have previously paid or incurred, but there
16   shall be no additional liability on its part to the other parties hereto by reason of such title failure;

17   (b) There shall be no retroactive adjustment of expenses incurred or revenues received from the operation of the
18   Lease or Interest which has failed, but the interests of the parties contained on Exhibit "A" shall be revised on an acreage
19   basis, as of the time it is determined finally that title failure has occurred, so that the interest of the party whose Lease or
20   Interest is affected by the title failure will thereafter be reduced in the Contract Area by the amount of the Lease or Interest failed;

21   (c) If the proportionate interest of the other parties hereto in any producing well previously drilled on the Contract
22   Area is increased by reason of the title failure, the party who bore the costs incurred in connection with such well attributable
23   to the Lease or Interest which has failed shall receive the proceeds attributable to the increase in such interest (less costs and
24   burdens attributable thereto) until it has been reimbursed for unrecovered costs paid by it in connection with such well
25   attributable to such failed Lease or Interest;

26   (d) Should any person not a party to this agreement, who is determined to be the owner of any Lease or Interest
27   which has failed, pay in any manner any part of the cost of operation, development, or equipment, such amount shall be paid
28   to the party or parties who bore the costs which are so refunded;

29   (e) Any liability to account to a person not a party to this agreement for prior production of Oil and Gas which arises
30   by reason of title failure shall be borne severally by each party (including a predecessor to a current party) who received
31   production for which such accounting is required based on the amount of such production received, and each such party shall
32   severally indemnify, defend and hold harmless all other parties hereto for any such liability to account;

33   (f) No charge shall be made to the joint account for legal expenses, fees or salaries in connection with the defense of
34   the Lease or Interest claimed to have failed, but if the party contributing such Lease or Interest hereto elects to defend its title
35   it shall bear all expenses in connection therewith; and

36   (g) If any party is given credit on Exhibit "A" to a Lease or Interest which is limited solely to ownership of an
37   interest in the wellbore of any well or wells and the production therefrom, such party's absence of interest in the remainder
38   of the Contract Area shall be considered a Failure of Title as to such remaining Contract Area unless that absence of interest
39   is reflected on Exhibit "A."

2. Loss by Non-Payment or Erroneous Payment of Amount Due: If, through mistake or oversight, any rental, shut-in well payment, minimum royalty or royalty payment, or other payment necessary to maintain all or a portion of an Oil and Gas Lease or Interest is not paid or is erroneously paid, and as a result a Lease or Interest terminates, there shall be no monetary liability against the party who failed to make such payment. Unless the party who failed to make the required payment secures a new Lease or Interest covering the same interest within ninety (90) days from the discovery of the failure to make proper payment, which acquisition will not be subject to Article VIII.B., the interests of the parties reflected on Exhibit "A" shall be revised on an acreage basis, effective as of the date of termination of the Lease or Interest involved, and the party who failed to make proper payment will no longer be credited with an interest in the Contract Area on account of ownership of the Lease or Interest which has terminated. If the party who failed to make the required payment shall not have been fully reimbursed, at the time of the loss, from the proceeds of the sale of Oil and Gas attributable to the lost Lease or Interest, calculated on an acreage basis, for the development and operating costs previously paid by it (but not for its share of the cost of any dry hole previously drilled or wells previously abandoned) from so much of the following as is necessary to effect reimbursement:

(a) Proceeds of Oil and Gas produced prior to termination of the Lease or Interest, less operating expenses and lease burdens chargeable hereunder to the person who failed to make payment, previously accrued to the credit of the lost Lease or Interest, on an acreage basis, up to the amount of unrecovered costs;

(b) Proceeds of Oil and Gas, less operating expenses and lease burdens chargeable hereunder to the person who failed to make payment, up to the amount of unrecovered costs attributable to that portion of Oil and Gas thereafter produced and marketed (excluding production from any wells thereafter drilled) which, in the absence of such Lease or Interest termination, would be attributable to the lost Lease or Interest on an acreage basis and which as a result of such Lease or Interest termination is credited to other parties, the proceeds of said portion of the Oil and Gas to be contributed by the other parties in proportion to their respective interests reflected on Exhibit "A"; and,

(c) Any monies, up to the amount of unrecovered costs, that may be paid by any party who is, or becomes, the owner of the Lease or Interest lost, for the privilege of participating in the Contract Area or becoming a party to this agreement.

3. Other Losses: All losses of Leases or Interests committed to this agreement, other than those set forth in Articles IV.B.1. and IV.B.2. above, shall be joint losses and shall be borne by all parties in proportion to their interests shown on Exhibit "A." This shall include but not be limited to the loss of any Lease or Interest through failure to develop or because express or implied covenants have not been performed (other than performance which requires only the payment of money), and the loss of any Lease by expiration at the end of its primary term if it is not renewed or extended. There shall be no readjustment of interests in the remaining portion of the Contract Area on account of any joint loss.

4. Curing Title: In the event of a Failure of Title under Article IV.B.1. or a loss of title under Article IV.B.2. above, any Lease or Interest acquired by any party hereto (other than the party whose interest has failed or was lost) during the ninety (90) day period provided by Article IV.B.1. and Article IV.B.2. above covering all or a portion of the interest that has failed or was lost shall be offered at cost to the party whose interest has failed or was lost, and the provisions of Article VIII.B. shall not apply to such acquisition.

- 3 -

A.A.P.L. FORM 610 - MODEL FORM OPERATING AGREEMENT - 1989

ARTICLE V.
OPERATOR

A. Designation and Responsibilities of Operator:

_____ shall be the Operator of the Contract Area, and shall conduct and direct and have full control of all operations on the Contract Area as permitted and required by, and within the limits of this agreement. In its performance of services hereunder for the Non-Operators, Operator shall be an independent contractor not subject to the control or direction of the Non-Operators except as to the type of operation to be undertaken in accordance with the election procedures contained in this agreement. Operator shall not be deemed, or hold itself out as, the agent of the Non-Operators with authority to bind them to any obligation or liability assumed or incurred by Operator as to any third party. Operator shall conduct its activities under this agreement as a reasonable prudent operator, in a good and workmanlike manner, with due diligence and dispatch, in accordance with good oilfield practice, and in compliance with applicable law and regulation, but in no event shall it have any liability as Operator to the other parties for losses sustained or liabilities incurred except such as may result from gross negligence or willful misconduct.

B. Resignation or Removal of Operator and Selection of Successor:

1. Resignation or Removal of Operator: Operator may resign at any time by giving written notice thereof to Non-Operators. If Operator terminates its legal existence, no longer owns an interest hereunder in the Contract Area, or is no longer capable of serving as Operator, Operator shall be deemed to have resigned without any action by Non-Operators, except the selection of a successor. Operator may be removed only for good cause by the affirmative vote of Non-Operators owning a majority interest based on ownership as shown on Exhibit "A" remaining after excluding the voting interest of Operator; such vote shall not be deemed effective until a written notice has been delivered to the Operator by a Non-Operator detailing the alleged default and Operator has failed to cure the default within thirty (30) days from its receipt of the notice or, if the default concerns an operation then being conducted, within forty-eight (48) hours of its receipt of the notice. For purposes hereof, "good cause" shall mean not only gross negligence or willful misconduct but also the material breach of or inability to meet the standards of operation contained in Article V.A. or material failure or inability to perform its obligations under this agreement.

Subject to Article VII.D.1., such resignation or removal shall not become effective until 7:00 o'clock A.M. on the first day of the calendar month following the expiration of ninety (90) days after the giving of notice of resignation by Operator or action by the Non-Operators to remove Operator, unless a successor Operator has been selected and assumes the duties of Operator at an earlier date. Operator, after effective date of resignation or removal, shall be bound by the terms hereof as a Non-Operator. A change of a corporate name or structure of Operator or transfer of Operator's interest to any single subsidiary, parent or successor corporation shall not be the basis for removal of Operator.

2. Selection of Successor Operator: Upon the resignation or removal of Operator under any provision of this agreement, a successor Operator shall be selected by the parties. The successor Operator shall be selected from the parties owning an interest in the Contract Area at the time such successor Operator is selected. The successor Operator shall be selected by the affirmative vote of two (2) or more parties owning a majority interest based on ownership as shown on Exhibit "A"; provided, however, if an Operator which has been removed or is deemed to have resigned fails to vote or votes only to succeed itself, the successor Operator shall be selected by the affirmative vote of the party or parties owning a majority interest based on ownership as shown on Exhibit "A" remaining after excluding the voting interest of the Operator that was removed or resigned. The former Operator shall promptly deliver to the successor Operator all records and data relating to the operations conducted by the former Operator to the extent such records and data are not already in the possession of the

40   successor operator. Any cost of obtaining or copying the former Operator's records and data shall be charged to the joint
41   account.
42   3. Effect of Bankruptcy: If Operator becomes insolvent, bankrupt or is placed in receivership, it shall be deemed to have
43   resigned without any action by Non-Operators, except the selection of a successor. If a petition for relief under the federal
44   bankruptcy laws is filed by or against Operator, and the removal of Operator is prevented by the federal bankruptcy court, all
45   Non-Operators and Operator shall comprise an interim operating committee to serve until Operator has elected to reject or
46   assume this agreement pursuant to the Bankruptcy Code, and an election to reject this agreement by Operator as a debtor in
47   possession, or by a trustee in bankruptcy, shall be deemed a resignation as Operator without any action by Non-Operators,
48   except the selection of a successor. During the period of time the operating committee controls operations, all actions shall
49   require the approval of two (2) or more parties owning a majority interest based on ownersh p as shown on Exhibit "A." In
50   the event there are only two (2) parties to this agreement, during the period of time the operating committee controls
51   operations, a third party acceptable to Operator, Non-Operator and the federal bankruptcy court shall be selected as a
52   member of the operating committee, and all actions shall require the approval of two (2) members of the operating
53   committee without regard for their interest in the Contract Area based on Exhibit "A."
54   **C. Employees and Contractors:**
55   The number of employees or contractors used by Operator in conducting operations hereunder, their selection, and the
56   hours of labor and the compensation for services performed shall be determined by Operator, and all such employees or
57   contractors shall be the employees or contractors of Operator.
58   **D. Rights and Duties of Operator:**
59   1. _Competitive Rates and Use of Affiliates:_ All wells drilled on the Contract Area shall be drilled on a competitive
60   contract basis at the usual rates prevailing in the area. If it so desires, Operator may employ its own tools and equipment in
61   the drilling of wells, but its charges therefor shall not exceed the prevailing rates in the area and the rate of such charges
62   shall be agreed upon by the parties in writing before drilling operations are commenced, and such work shall be performed by
63   Operator under the same terms and conditions as are customary and usual in the area in contracts of independent contractors
64   who are doing work of a similar nature. All work performed or materials supplied by affiliates or related parties of Operator
65   shall be performed or supplied at competitive rates, pursuant to written agreement, and in accordance with customs and
66   standards prevailing in the industry.
67   2. Discharge of Joint Account Obligations: Except as herein otherwise specifically provided, Operator shall promptly pay
68   and discharge expenses incurred in the development and operation of the Contract Area pursuant to this agreement and shall
69   charge each of the parties hereto with their respective proportionate shares upon the expense basis provided in Exhibit "C."
70   Operator shall keep an accurate record of the joint account hereunder, showing expenses incurred and charges and credits
71   made and received.
72   3. Protection from Liens: Operator shall pay, or cause to be paid, as and when they become due and payable, all accounts
73   of contractors and suppliers and wages and salaries for services rendered or performed, and for materials supplied on, to or in
74   respect of the Contract Area or any operations for the joint account thereof, and shall keep the Contract Area free from

- 4 -

A.A.P.L. FORM 610 - MODEL FORM OPERATING AGREEMENT - 1989

1.   liens and encumbrances resulting therefrom except for those resulting from a bona fide dispute as to services rendered or
2.   materials supplied.
3.   4. Custody of Funds: Operator shall hold for the account of the Non-Operators any funds of the Non-Operators advanced
4.   or paid to the Operator, either for the conduct of operations hereunder or as a result of the sale of production from the
5.   Contract Area, and such funds shall remain the funds of the Non-Operators on whose account they are advanced or paid until
6.   used for their intended purpose or otherwise delivered to the Non-Operators or applied toward the payment of debts as
7.   provided in Article VII.B. Nothing in this paragraph shall be construed to establish a fiduciary relationship between Operator
8.   and Non-Operators for any purpose other than to account for Non-Operator funds as herein specifically provided. Nothing in
9.   this paragraph shall require the maintenance by Operator of separate accounts for the funds of Non-Operators unless the
10.   parties otherwise specifically agree.
11.   5. Access to Contract Area and Records: Operator shall, except as otherwise provided herein, permit each Non-Operator
12.   or its duly authorized representative, at the Non-Operator's sole risk and cost, full and free access at all reasonable times to
13.   all operations of every kind and character being conducted for the joint account on the Contract Area and to the records of
14.   operations conducted thereon or production therefrom, including Operator's books and records relating thereto. Such access
15.   rights shall not be exercised in a manner interfering with Operator's conduct of an operation hereunder and shall not obligate
16.   Operator to furnish any geologic or geophysical data of an interpretive nature unless the cost of preparation of such
17.   interpretive data was charged to the joint account. Operator will furnish to each Non-Operator upon request copies of any
18.   and all reports and information obtained by Operator in connection with production and related items, including, without
19.   limitation, meter and chart reports, production purchaser statements, run tickets and monthly gauge reports, but excluding
20.   purchase contracts and pricing information to the extent not applicable to the production of the Non-Operator seeking the
21.   information. Any audit of Operator's records relating to amounts expended and the appropriateness of such expenditures
22.   shall be conducted in accordance with the audit protocol specified in Exhibit "C."
23.   6. Filing and Furnishing Governmental Reports: Operator will file, and upon written request promptly furnish copies to
24.   each requesting Non-Operator not in default of its payment obligations, all operational notices, reports or applications
25.   required to be filed by local, State, Federal or Indian agencies or authorities having jurisdiction over operations hereunder.
26.   Each Non-Operator shall provide to Operator on a timely basis all information necessary to Operator to make such filings.
27.   7. Drilling and Testing Operations: The following provisions shall apply to each well drilled hereunder, including but not
28.   limited to the Initial Well:
29.   (a) Operator will promptly advise Non-Operators of the date on which the well is spudded, or the date on which
30.   drilling operations are commenced.
31.   (b) Operator will send to Non-Operators such reports, test results and notices regarding the progress of operations on the well
32.   as the Non-Operators shall reasonably request, including, but not limited to, daily drilling reports, completion reports, and well logs.
33.   (c) Operator shall adequately test all Zones encountered which may reasonably be expected to be capable of producing
34.   Oil and Gas in paying quantities as a result of examination of the electric log or any other logs or cores or tests conducted
35.   hereunder.

36 8. Cost Estimates. Upon request of any Consenting Party, Operator shall furnish estimates of current and cumulative costs
37 incurred for the joint account at reasonable intervals during the conduct of any operation pursuant to this agreement.
38 Operator shall not be held liable for errors in such estimates so long as the estimates are made in good faith.
39 9. Insurance: At all times while operations are conducted hereunder, Operator shall comply with the workers
40 compensation law of the state where the operations are being conducted; provided, however, that Operator may be a self-
41 insurer for liability under said compensation laws in which event the only charge that shall be made to the joint account shall
42 be as provided in Exhibit "C." Operator shall also carry or provide insurance for the benefit of the joint account of the parties
43 as outlined in Exhibit "D" attached hereto and made a part hereof. Operator shall require all contractors engaged in work on
44 or for the Contract Area to comply with the workers compensation law of the state where the operations are being conducted
45 and to maintain such other insurance as Operator may require.
46 In the event automobile liability insurance is specified in said Exhibit "D," or subsequently receives the approval of the
47 parties, no direct charge shall be made by Operator for premiums paid for such insurance for Operator's automotive
48 equipment.
49
50 ARTICLE VI.
DRILLING AND DEVELOPMENT
51 A. Initial Well:
52 On or before the _____ day of _____, 19 ____, Operator shall commence the drilling of the Initial
53 Well at the following location:
54
55
56
57
58
59
60 and shall thereafter continue the drilling of the well with due diligence to
61
62
63
64
65
66
67 The drilling of the Initial Well and the participation therein by all parties is obligatory, subject to Article VI.C.1. as to participation
68 in Completion operations and Article VI.F. as to termination of operations and Article XI as to occurrence of force majeure.
69 B. Subsequent Operations:
70 1. Proposed Operations: If any party hereto should desire to drill any well on the Contract Area other than the Initial Well, or
71 if any party should desire to Rework, Sidetrack, Deepen, Recomplete or Plug Back a dry hole or a well no longer capable of
72 producing in paying quantities in which such party has not otherwise relinquished its interest in the proposed objective Zone under
73 this agreement, the party desiring to drill, Rework, Sidetrack, Deepen, Recomplete or Plug Back such a well shall give written
74 notice of the proposed operation to the parties who have not otherwise relinquished their interest in such objective Zone

- 5 -

213

A.A.P.L. FORM 610 - MODEL FORM OPERATING AGREEMENT - 1989

1   under this agreement and to all other parties in the case of a proposal for Sidetracking or Deepening, specifying the work to be
2   performed, the location, proposed depth, objective Zone and the estimated cost of the operation. The parties to whom such a
3   notice is delivered shall have thirty (30) days after receipt of the notice within which to notify the party proposing to do the work
4   whether they elect to participate in the cost of the proposed operation. If a drilling rig is on location, notice of a proposal to
5   Rework, Sidetrack, Recomplete, Plug Back or Deepen may be given by telephone and the response period shall be limited to forty-
6   eight (48) hours, exclusive of Saturday, Sunday and legal holidays. Failure of a party to whom such notice is delivered to reply
7   within the period above fixed shall constitute an election by that party not to participate in the cost of the proposed operation.
8   Any proposal by a party to conduct an operation conflicting with the operation initially proposed shall be delivered to all parties
9   within the time and in the manner provided in Article VI.B.6.

10   If all parties to whom such notice is delivered elect to participate in such a proposed operation, the parties shall be
11   contractually committed to participate therein provided such operations are commenced within the time period hereafter set
12   forth, and Operator shall, no later than ninety (90) days after expiration of the notice period of thirty (30) days (or as
13   promptly as practicable after the expiration of the forty-eight (48) hour period when a drilling rig is on location, as the case
14   may be), actually commence the proposed operation and thereafter complete it with due diligence at the risk and expense of
15   the parties participating therein; provided, however, said commencement date may be extended upon written notice of same
16   by Operator to the other parties, for a period of up to thirty (30) additional days if, in the sole opinion of Operator, such
17   additional time is reasonably necessary to obtain permits from governmental authorities, surface rights (including rights-of-
18   way) or appropriate drilling equipment, or to complete title examination or curative matter required for title approval or
19   acceptance. If the actual operation has not been commenced within the time provided (including any extension thereof as
20   specifically permitted herein or in the force majeure provisions of Article XI) and if any party hereto still desires to conduct
21   said operation, written notice proposing same must be resubmitted to the other parties in accordance herewith as if no prior
22   proposal had been made. Those parties that did not participate in the drilling of a well for which a proposal to Deepen or
23   Sidetrack is made hereunder shall, if such parties desire to participate in the proposed Deepening or Sidetracking operation,
24   reimburse the Drilling Parties in accordance with Article VI.B.4. in the event of a Deepening operation and in accordance
25   with Article VI.B.5. in the event of a Sidetracking operation.

26   2. Operations by Less Than All Parties:

27   (a) Determination of Participation. If any party to whom such notice is delivered as provided in Article VI.B.1. or
28   VI.C.1. (Option No. 2) elects not to participate in the proposed operation, then, in order to be entitled to the benefits of this
29   Article, the party or parties giving the notice and such other parties as shall elect to participate in the operation shall, no
30   later than ninety (90) days after the expiration of the notice period of thirty (30) days (or as promptly as practicable after the
31   expiration of the forty-eight (48) hour period when a drilling rig is on location, as the case may be) actually commence the
32   proposed operation and complete it with due diligence. Operator shall perform all work for the account of the Consenting
33   Parties; provided, however, if no drilling rig or other equipment is on location, and if Operator is a Non-Consenting Party,
34   the Consenting Parties shall either: (i) request Operator to perform the work required by such proposed operation for the
35   account of the Consenting Parties, or (ii) designate one of the Consenting Parties as Operator to perform such work. The
36   rights and duties granted to and imposed upon the Operator under this agreement are granted to and imposed upon the party
37   designated as Operator for an operation in which the original Operator is a Non-Consenting Party. Consenting Parties, when
38   conducting operations on the Contract Area pursuant to this Article VI.B.2, shall comply with all terms and conditions of this
39   agreement.

40   If less than all parties approve any proposed operation, the proposing party, immediately after the expiration of the
41 applicable notice period, shall advise all Parties of the total interest of the parties approving such operation and its
42 recommendation as to whether the Consenting Parties should proceed with the operation as proposed. Each Consenting Party,
43 within forty-eight (48) hours (exclusive of Saturday, Sunday and legal holidays) after delivery of such notice, shall advise the
44 proposing party of its desire to (i) limit participation to such party's interest as shown on Exhibit "A" or (ii) carry only its
45 proportionate part (determined by dividing such party's interest in the Contract Area by the interests of all Consenting Parties in
46 the Contract Area) of Non-Consenting Parties' interests, or (iii) carry its proportionate part (determined as provided in (iii)) of
47 Non-Consenting Parties' interests together with all or a portion of its proportionate part of any Non-Consenting Parties'
48 interests that any Consenting Party did not elect to take. Any interest of Non-Consenting Parties that is not carried by a
49 Consenting Party shall be deemed to be carried by the party proposing the operation if such party does not withdraw its
50 proposal. Failure to advise the proposing party within the time required shall be deemed an election under (i) . In the event a
51 drilling rig is on location, notice may be given by telephone, and the time permitted for such a response shall not exceed a
52 total of forty-eight (48) hours (exclusive of Saturday, Sunday and legal holidays). The proposing party, at its election, may
53 withdraw such proposal if there is less than 100% participation and shall notify all parties of such decision within ten (10)
54 days, or within twenty-four (24) hours if a drilling rig is on location, following expiration of the applicable response period.
55 If 100% subscription to the proposed operation is obtained, the proposing party shall promptly notify the Consenting Parties
56 of their proportionate interests in the operation and the party serving as Operator shall commence such operation within the
57 period provided in Article VI.B.1., subject to the same extension right as provided therein.
58   (b) Relinquishment of Interest for Non-Participation. The entire cost and risk of conducting such operations shall be
59 borne by the Consenting Parties in the proportions they have elected to bear same under the terms of the preceding
60 paragraph. Consenting Parties shall keep the leasehold estates involved in such operations free and clear of all liens and
61 encumbrances of every kind created by or arising from the operations of the Consenting Parties. If such an operation results
62 in a dry hole, then subject to Articles VI.B.6. and VI.E.3., the Consenting Parties shall plug and abandon the well and restore
63 the surface location at their sole cost, risk and expense; provided, however, that those Non-Consenting Parties that
64 participated in the drilling, Deepening or Sidetracking of the well shall remain liable for, and shall pay, their proportionate
65 shares of the cost of plugging and abandoning the well and restoring the surface location insofar only as those costs were not
66 increased by the subsequent operations of the Consenting Parties. If any well drilled, Reworked, Sidetracked, Deepened,
67 Recompleted or Plugged Back under the provisions of this Article results in a well capable of producing Oil and/or Gas in
68 paying quantities, the Consenting Parties shall Complete and equip the well to produce at their sole cost and risk, and the
69 well shall then be turned over to Operator (if the Operator did not conduct the operation) and shall be operated by it at the
70 expense and for the account of the Consenting Parties. Upon commencement of operations for the drilling, Reworking,
71 Sidetracking, Recompleting, Deepening or Plugging Back of any such well by Consenting Parties in accordance with the
72 provisions of this Article, each Non-Consenting Party shall be deemed to have relinquished to Consenting Parties, and the
73 Consenting Parties shall own and be entitled to receive, in proportion to their respective interests, all of such Non-
74 Consenting Party's interest in the well and share of production therefrom or, in the case of a Reworking, Sidetracking,

- 6 -

215

A.A.P.L. FORM 610 - MODEL FORM OPERATING AGREEMENT - 1989

1 Deepening, Recompleting or Plugging Back, or a Completion pursuant to Article VI.C.1. Option No. 2, all of such Non-
2 Consenting Party's interest in the production obtained from the operation in which the Non-Consenting Party did not elect
3 to participate. Such relinquishment shall be effective until the proceeds of the sale of such share, calculated at the well, or
4 market value thereof if such share is not sold (after deducting applicable ad valorem, production, severance, and excise taxes,
5 royalty, overriding royalty and other interests not excepted by Article III.C. payable out of or measured by the production
6 from such well accruing with respect to such interest until it reverts), shall equal the total of the following:

7 (i) _____ % of each such Non-Consenting Party's share of the cost of any newly acquired surface equipment
8 beyond the wellhead connections (including but not limited to stock tanks, separators, treaters, pumping equipment and
9 piping), plus 100% of each such Non-Consenting Party's share of the cost of operation of the well commencing with first
10 production and continuing until each such Non-Consenting Party's relinquished interest shall revert to it under other
11 provisions of this Article, it being agreed that each Non-Consenting Party's share of such costs and equipment will be that
12 interest which would have been chargeable to such Non-Consenting Party had it participated in the well from the beginning
13 of the operations; and

14 (ii) _____ % of (a) that portion of the costs and expenses of drilling, Reworking, Sidetracking, Deepening,
15 Plugging Back, testing, Completing, and Recompleting, after deducting any cash contributions received under Article VIII.C,
16 and of (b) that portion of the cost of newly acquired equipment in the well (to and including the wellhead connections),
17 which would have been chargeable to such Non-Consenting Party if it had participated therein.

18 Notwithstanding anything to the contrary in this Article VI.B, if the well does not reach the deepest objective Zone
19 described in the notice proposing the well for reasons other than the encountering of granite or practically impenetrable
20 substance or other condition in the hole rendering further operations impracticable, Operator shall give notice thereof to each
21 Non-Consenting Party who submitted or voted for an alternative proposal under Article VI.B.6. to drill the well to a
22 shallower Zone than the deepest objective Zone proposed in the notice under which the well was drilled, and each such Non-
23 Consenting Party shall have the option to participate in the initial proposed Completion of the well by paying its share of the
24 cost of drilling the well to its actual depth, calculated in the manner provided in Article VI.B.4. (a). If any such Non-
25 Consenting Party does not elect to participate in the first Completion proposed for such well, the relinquishment provisions
26 of this Article VI.B.2. (b) shall apply to such party's interest.

27 (c) Reworking, Recompleting or Plugging Back. An election not to participate in the drilling, Sidetracking or
28 Deepening of a well shall be deemed an election not to participate in any Reworking or Plugging Back operation proposed in
29 such a well, or portion thereof, to which the initial non-consent election applied that is conducted at any time prior to full
30 recovery by the Consenting Parties of the Non-Consenting Party's recoupment amount. Similarly, an election not to
31 participate in the Completing or Recompleting of a well shall be deemed an election not to participate in any Reworking
32 operation proposed in such a well, or portion thereof, to which the initial non-consent election applied that is conducted at
33 any time prior to full recovery by the Consenting Parties of the Non-Consenting Party's recoupment amount. Any such
34 Reworking, Recompleting or Plugging Back operation conducted during the recoupment period shall be deemed part of the
35 cost of operation of said well and there shall be added to the sums to be recouped by the Consenting Parties _____ % of
36 that portion of the costs of the Reworking, Recompleting or Plugging Back operation which would have been chargeable to
37 such Non-Consenting Party had it participated therein. If such a Reworking, Recompleting or Plugging Back operation is
38 proposed during such recoupment period, the provisions of this Article VI.B. shall be applicable as between said Consenting
39 Parties in said well.

40   (d) Recoupment Matters. During the period of time Consenting Parties are entitled to receive Non-Consenting Party's
41 share of production, or the proceeds therefrom, Consenting Parties shall be responsible for the payment of all ad valorem,
42 production, severance, excise, gathering and other taxes, and all royalty, overriding royalty and other burdens applicable to
43 Non-Consenting Party's share of production not excepted by Article III.C.
44   In the case of any Reworking, Sidetracking, Plugging Back, Recompleting or Deepening operation, the Consenting
45 Parties shall be permitted to use, free of cost, all casing, tubing and other equipment in the well, but the ownership of all
46 such equipment shall remain unchanged; and upon abandonment of a well after such Reworking, Sidetracking, Plugging Back,
47 Recompleting or Deepening, the Consenting Parties shall account for all such equipment to the owners thereof, with each
48 party receiving its proportionate part in kind or in value, less cost of salvage.
49   Within ninety (90) days after the completion of any operation under this Article, the party conducting the operations
50 for the Consenting Parties shall furnish each Non-Consenting Party with an inventory of the equipment in and connected to
51 the well, and an itemized statement of the cost of drilling, Sidetracking, Deepening, Plugging Back, testing, Completing,
52 Recompleting, and equipping the well for production; or, at its option, the operating party, in lieu of an itemized statement
53 of such costs of operation, may submit a detailed statement of monthly billings. Each month thereafter, during the time the
54 Consenting Parties are being reimbursed as provided above, the party conducting the operations for the Consenting Parties
55 shall furnish the Non-Consenting Parties with an itemized statement of all costs and liabilities incurred in the operation of
56 the well, together with a statement of the quantity of Oil and Gas produced from it and the amount of proceeds realized from
57 the sale of the well's working interest production during the preceding month. In determining the quantity of Oil and Gas
58 produced during any month, Consenting Parties shall use industry accepted methods such as but not limited to metering or
59 periodic well tests. Any amount realized from the sale or other disposition of equipment newly acquired in connection with
60 any such operation which would have been owned by a Non-Consenting Party had it participated therein shall be credited
61 against the total unreturned costs of the work done and of the equipment purchased in determining when the interest of such
62 Non-Consenting Party shall revert to it as above provided; and if there is a credit balance, it shall be paid to such Non-
63 Consenting Party.
64   If and when the Consenting Parties recover from a Non-Consenting Party's relinquished interest the amounts provided
65 for above, the relinquished interests of such Non-Consenting Party shall automatically revert to it as of 7:00 a.m. on the day
66 following the day on which such recoupment occurs, and, from and after such reversion, such Non-Consenting Party shall
67 own the same interest in such well, the material and equipment in or pertaining thereto, and the production therefrom as
68 such Non-Consenting Party would have been entitled to had it participated in the drilling, Sidetracking, Reworking,
69 Deepening, Recompleting or Plugging Back of said well. Thereafter, such Non-Consenting Party shall be charged with and
70 shall pay its proportionate part of the further costs of the operation of said well in accordance with the terms of this
71 agreement and Exhibit "C" attached hereto.
72   3. Stand-By Costs: When a well which has been drilled or Deepened has reached its authorized depth and all tests have
73 been completed and the results thereof furnished to the parties, or when operations on the well have been otherwise
74 terminated pursuant to Article VI.F, stand-by costs incurred pending response to a party's notice proposing a Reworking,

A.A.P.L. FORM 610 - MODEL FORM OPERATING AGREEMENT - 1989

1   Sidetracking, Deepening, Recompleting, Plugging Back or Completing operation in such a well (including the period required
2   under Article VI.B.6. to resolve competing proposals) shall be charged and borne as part of the drilling or Deepening
3   operation just completed. Stand-by costs subsequent to all parties responding, or expiration of the response time permitted,
4   whichever first occurs, and prior to agreement as to the participating interests of all Consenting Parties pursuant to the terms
5   of the second grammatical paragraph of Article VI.B.2. (a), shall be charged to and borne as part of the proposed operation,
6   but if the proposal is subsequently withdrawn because of insufficient participation, such stand-by costs shall be allocated
7   between the Consenting Parties in the proportion each Consenting Party's interest as shown on Exhibit "A" bears to the total
8   interest as shown on Exhibit "A" of all Consenting Parties.

9   In the event that notice for a Sidetracking operation is given while the drilling rig to be utilized is on location, any party
10   may request and receive up to five (5) additional days after expiration of the forty-eight hour response period specified in
11   Article VI.B.1. within which to respond by paying for all stand-by costs and other costs incurred during such extended
12   response period; Operator may require such party to pay the estimated stand-by time in advance as a condition to extending
13   the response period. If more than one party elects to take such additional time to respond to the notice, standby costs shall be
14   allocated between the parties taking additional time to respond on a day-to-day basis in the proportion each electing party's
15   interest as shown on Exhibit "A" bears to the total interest as shown on Exhibit "A" of all the electing parties.

16   4. **Deepening:** If less than all the parties elect to participate in a drilling, Sidetracking, or Deepening operation proposed
17   pursuant to Article VI.B.1., the interest relinquished by the Non-Consenting Parties to the Consenting Parties under Article
18   VI.B.2. shall relate only and be limited to the lesser of (i) the total depth actually drilled or (ii) the objective depth or Zone
19   of which the parties were given notice under Article VI.B.1. ("Initial Objective"). Such well shall not be Deepened beyond the
20   Initial Objective without first complying with this Article to afford the Non-Consenting Parties the opportunity to participate
21   in the Deepening operation.

22   In the event any Consenting Party desires to drill or Deepen a Non-Consent Well to a depth below the Initial Objective,
23   such party shall give notice thereof, complying with the requirements of Article VI.B.1., to all parties (including Non-
24   Consenting Parties). Thereupon, Articles VI.B.1. and 2. shall apply and all parties receiving such notice shall have the right to
25   participate or not participate in the Deepening of such well pursuant to said Articles VI.B.1. and 2. If a Deepening operation
26   is approved pursuant to such provisions, and if any Non-Consenting Party elects to participate in the Deepening operation,
27   such Non-Consenting party shall pay or make reimbursement (as the case may be) of the following costs and expenses:

28   (a) If the proposal to Deepen is made prior to the Completion of such well as a well capable of producing in paying
29   quantities, such Non-Consenting Party shall pay (or reimburse Consenting Parties for, as the case may be) that share of costs
30   and expenses incurred in connection with the drilling of said well from the surface to the Initial Objective which Non-
31   Consenting Party would have paid had such Non-Consenting Party agreed to participate therein, plus the Non-Consenting
32   Party's share of the cost of Deepening and of participating in any further operations on the well in accordance with the other
33   provisions of this Agreement; provided, however, all costs for testing and Completion or attempted Completion of the well
34   incurred by Consenting Parties prior to the point of actual operations to Deepen beyond the Initial Objective shall be for the
35   sole account of Consenting Parties.

36   (b) If the proposal is made for a Non-Consent Well that has been previously Completed as a well capable of producing
37   in paying quantities, but is no longer capable of producing in paying quantities, such Non-Consenting Party shall pay (or

38 reimburse Consenting Parties for, as the case may be) its proportionate share of all costs of drilling, Completing, and
39 equipping said well from the surface to the Initial Objective, calculated in the manner provided in paragraph (a) above, less
40 those costs recouped by the Consenting Parties from the sale of production from the well. The Non-Consenting Party shall
41 also pay its proportionate share of all costs of re-entering said well. The Non-Consenting Parties proportionate part (based
42 on the percentage of such well Non-Consenting Party would have owned had it previously participated in such Non-Consent
43 Well) of the costs of salvable materials and equipment remaining in the hole and salvable surface equipment used in
44 connection with such well shall be determined in accordance with Exhibit "C." If the Consenting Parties have recouped the
45 cost of drilling, Completing, and equipping the well at the time such Deepening operation is conducted, then a Non-
46 Consenting Party may participate in the Deepening of the well with no payment for costs incurred prior to re-entering the
47 well for Deepening.
48   The foregoing shall not imply a right of any Consenting Party to propose any Deepening for a Non-Consent Well prior
49 to the drilling of such well to its Initial Objective without the consent of the other Consenting Parties as provided in Article
50 VI.F.
51   5. Sidetracking: Any party having the right to participate in a proposed Sidetracking operation that does not own an
52 interest in the affected wellbore at the time of the notice shall, upon electing to participate, tender to the wellbore owners its
53 proportionate share (equal to its interest in the Sidetracking operation) of the value of that portion of the existing wellbore
54 to be utilized as follows:
55   (a) If the proposal is for Sidetracking an existing dry hole, reimbursement shall be on the basis of the actual costs
56 incurred in the initial drilling of the well down to the depth at which the Sidetracking operation is initiated.
57   (b) If the proposal is for Sidetracking a well which has previously produced, reimbursement shall be on the basis of
58 such party's proportionate share of drilling and equipping costs incurred in the initial drilling of the well down to the depth
59 at which the Sidetracking operation is conducted, calculated in the manner described in Article VI.B.4(b) above. Such party's
60 proportionate share of the cost of the well's salvable materials and equipment down to the depth at which the Sidetracking
61 operation is initiated shall be determined in accordance with the provisions of Exhibit "C."
62   6. Order of Preference of Operations. Except as otherwise specifically provided in this agreement, if any party desires to
63 propose the conduct of an operation that conflicts with a proposal that has been made by a party under this Article VI, such
64 party shall have fifteen (15) days from delivery of the initial proposal, in the case of a proposal to drill a well or to perform
65 an operation on a well where no drilling rig is on location, or twenty-four (24) hours, exclusive of Saturday, Sunday and legal
66 holidays, from delivery of the initial proposal, if a drilling rig is on location for the well on which such operation is to be
67 conducted, to deliver to all parties entitled to participate in the proposed operation such party's alternative proposal, such
68 alternate proposal to contain the same information required to be included in the initial proposal. Each party receiving such
69 proposals shall elect by delivery of notice to Operator within five (5) days after expiration of the proposal period, or within
70 twenty-four (24) hours (exclusive of Saturday, Sunday and legal holidays) if a drilling rig is on location for the well that is the
71 subject of the proposals, to participate in one of the competing proposals. Any party not electing within the time required
72 shall be deemed not to have voted. The proposal receiving the vote of parties owning the largest aggregate percentage
73 interest of the parties voting shall have priority over all other competing proposals; in the case of a tie vote, the

- 8 -

A.A.P.L. FORM 610 - MODEL FORM OPERATING AGREEMENT - 1989

1   initial proposal shall prevail. Operator shall deliver notice of such result to all parties entitled to participate in the operation
2   within five (5) days after expiration of the election period (or within twenty-four (24) hours, exclusive of Saturday, Sunday
3   and legal holidays, if a drilling rig is on location). Each party shall then have two (2) days (or twenty-four (24) hours if a rig
4   is on location) from receipt of such notice to elect by delivery of notice to Operator to participate in such operation or to
5   relinquish interest in the affected well pursuant to the provisions of Article VI.B.2.; failure by a party to deliver notice within
6   such period shall be deemed an election not to participate in the prevailing proposal.

7   7. Conformity to Spacing Pattern. Notwithstanding the provisions of this Article VI.B.2., it is agreed that no wells shall be
8   proposed to be drilled to or Completed in or produced from a Zone from which a well located elsewhere on the Contract
9   Area is producing, unless such well conforms to the then-existing well spacing pattern for such Zone.

10   8. Paying Wells. No party shall conduct any Reworking, Deepening, Plugging Back, Completion, Recompletion, or
11   Sidetracking operation under this agreement with respect to any well then capable of producing in paying quantities except
12   with the consent of all parties that have not relinquished interests in the well at the time of such operation.

13   **C. Completion of Wells; Reworking and Plugging Back:**

14   1. Completion: Without the consent of all parties, no well shall be drilled, Deepened or Sidetracked, except any well
15   drilled, Deepened or Sidetracked pursuant to the provisions of Article VI.B.2. of this agreement. Consent to the drilling,
16   Deepening or Sidetracking shall include:

17   ☐ Option No. 1: All necessary expenditures for the drilling, Deepening or Sidetracking, testing, Completing and
18   equipping of the well, including necessary tankage and/or surface facilities.

19   ☐ Option No. 2: All necessary expenditures for the drilling, Deepening or Sidetracking and testing of the well. When
20   such well has reached its authorized depth, and all logs, cores and other tests have been completed, and the results
21   thereof furnished to the parties, Operator shall give immediate notice to the Non-Operators having the right to
22   participate in a Completion attempt whether or not Operator recommends attempting to Complete the well,
23   together with Operator's AFE for Completion costs if not previously provided. The parties receiving such notice
24   shall have forty-eight (48) hours (exclusive of Saturday, Sunday and legal holidays) in which to elect by delivery of
25   notice to Operator to participate in a recommended Completion attempt or to make a Completion proposal with an
26   accompanying AFE. Operator shall deliver any such Completion proposal, or any Completion proposal conflicting
27   with Operator's proposal, to the other parties entitled to participate in such Completion in accordance with the
28   procedures specified in Article VI.B.6. Election to participate in a Completion attempt shall include consent to all
29   necessary expenditures for the Completing and equipping of such well, including necessary tankage and/or surface
30   facilities but excluding any stimulation operation not contained on the Completion AFE. Failure of any party
31   receiving such notice to reply within the period above fixed shall constitute an election by that party not to
32   participate in the cost of the Completion attempt; provided, that Article VI.B.6. shall control in the case of
33   conflicting Completion proposals. If one or more, but less than all of the parties, elect to attempt a Completion, the
34   provisions of Article VI.B.2. hereof (the phrase "Reworking, Sidetracking, Deepening, Recompleting or Plugging
35   Back" as contained in Article VI.B.2. shall be deemed to include "Completing") shall apply to the operations
36   thereafter conducted by less than all parties; provided, however, that Article VI.B.2 shall apply separately to each
37   separate Completion or Recompletion attempt undertaken hereunder, and an election to become a Non-Consenting

38   Party as to one Completion or Recompletion attempt shall not prevent a party from becoming a Consenting Party

39 in subsequent Completion or Recompletion attempts regardless whether the Consenting Parties as to earlier

40 Completions or Recompletions have recouped their costs pursuant to Article VI.B.2.; provided further, that any

41 recoupment of costs by a Consenting Party shall be made solely from the production attributable to the Zone in

42 which the Completion attempt is made. Election by a previous Non-Consenting Party to participate in a subsequent

43 Completion or Recompletion attempt shall require such party to pay its proportionate share of the cost of salvable

44 materials and equipment installed in the well pursuant to the previous Completion or Recompletion attempt,

45 insofar and only insofar as such materials and equipment benefit the Zone in which such party participates in a

46 Completion attempt.

47   2. Rework, Recomplete or Plug Back: No well shall be Reworked, Recompleted or Plugged Back except a well Reworked,

48 Recompleted, or Plugged Back pursuant to the provisions of Article VI.B.2. of this agreement. Consent to the Reworking,

49 Recompleting or Plugging Back of a well shall include all necessary expenditures in conducting such operations and

50 Completing and equipping of said well, including necessary tankage and/or surface facilities.

51 **D. Other Operations:**

52   Operator shall not undertake any single project reasonably estimated to require an expenditure in excess of _____

53 _____ Dollars ($ _____) except in connection with the

54 drilling, Sidetracking, Reworking, Deepening, Completing, Recompleting or Plugging Back of a well that has been previously

55 authorized by or pursuant to this agreement; provided, however, that, in case of explosion, fire, flood or other sudden

56 emergency, whether of the same or different nature, Operator may take such steps and incur such expenses as in its opinion

57 are required to deal with the emergency to safeguard life and property but Operator, as promptly as possible, shall report the

58 emergency to the other parties. If Operator prepares an AFE for its own use, Operator shall furnish any Non-Operator so

59 requesting an information copy thereof for any single project costing in excess of _____ Dollars

60 ($ _____). Any party who has not relinquished its interest in a well shall have the right to propose that

61 Operator perform repair work or undertake the installation of artificial lift equipment or ancillary production facilities such as

62 salt water disposal wells or to conduct additional work with respect to a well drilled hereunder or other similar project (but

63 not including the installation of gathering lines or other transportation or marketing facilities, the installation of which shall

64 be governed by separate agreement between the parties) reasonably estimated to require an expenditure in excess of the

65 amount first set forth above in this Article VI.D. (except in connection with an operation required to be proposed under

66 Articles VI.B.1. or VI.C.1. Option No. 2, which shall be governed exclusively by those Articles). Operator shall deliver such

67 proposal to all parties entitled to participate therein. If within thirty (30) days thereof Operator secures the written consent

68 of any party or parties owning at least _____ % of the interests of the parties entitled to participate in such operation,

69 each party having the right to participate in such project shall be bound by the terms of such proposal and shall be obligated

70 to pay its proportionate share of the costs of the proposed project as if it had consented to such project pursuant to the terms

71 of the proposal.

72 **E. Abandonment of Wells:**

73   1. Abandonment of Dry Holes: Except for any well drilled or Deepened pursuant to Article VI.B.2., any well which has

74 been drilled or Deepened under the terms of this agreement and is proposed to be completed as a dry hole shall not be

- 9 -

A.A.P.L. FORM 610 - MODEL FORM OPERATING AGREEMENT - 1989

1   plugged and abandoned without the consent of all parties. Should Operator, after diligent effort, be unable to contact any
2   party, or should any party fail to reply within forty-eight (48) hours (exclusive of Saturday, Sunday and legal holidays) after
3   delivery of notice of the proposal to plug and abandon such well, such party shall be deemed to have consented to the
4   proposed abandonment. All such wells shall be plugged and abandoned in accordance with applicable regulations and at the
5   cost, risk and expense of the parties who participated in the cost of drilling or Deepening such well. Any party who objects to
6   plugging and abandoning such well by notice delivered to Operator within forty-eight (48) hours (exclusive of Saturday,
7   Sunday and legal holidays) after delivery of notice of the proposed plugging shall take over the well as of the end of such
8   forty-eight (48) hour notice period and conduct further operations in search of Oil and/or Gas subject to the provisions of
9   Article VI.B.; failure of such party to provide proof reasonably satisfactory to Operator of its financial capability to conduct
10 such operations or to take over the well within such period or thereafter to conduct operations on such well or plug and
11 abandon such well shall entitle Operator to retain or take possession of the well and plug and abandon the well. The party
12 taking over the well shall indemnify Operator (if Operator is an abandoning party) and the other abandoning parties against
13 liability for any further operations conducted on such well except for the costs of plugging and abandoning the well and
14 restoring the surface, for which the abandoning parties shall remain proportionately liable.

15   2. Abandonment of Wells That Have Produced: Except for any well in which a Non-Consent operation has been
16 conducted hereunder for which the Consenting Parties have not been fully reimbursed as herein provided, any well which has
17 been completed as a producer shall not be plugged and abandoned without the consent of all parties. If all parties consent to
18 such abandonment, the well shall be plugged and abandoned in accordance with applicable regulations and at the cost, risk
19 and expense of all the parties hereto. Failure of a party to reply within sixty (60) days of delivery of notice of proposed
20 abandonment shall be deemed an election to consent to the proposal. If, within sixty (60) days after delivery of notice of the
21 proposed abandonment of any well, all parties do not agree to the abandonment of such well, those wishing to continue its
22 operation from the Zone then open to production shall be obligated to take over the well as of the expiration of the
23 applicable notice period and shall indemnify Operator (if Operator is an abandoning party) and the other abandoning parties
24 against liability for any further operations on the well conducted by such parties. Failure of such party or parties to provide
25 proof reasonably satisfactory to Operator of their financial capability to conduct such operations or to take over the well
26 within the required period or thereafter to conduct operations on such well shall entitle Operator to retain or take possession
27 of such well and plug and abandon the well.

28   Parties taking over a well as provided herein shall tender to each of the other parties its proportionate share of the value of
29 the well's salvable material and equipment, determined in accordance with the provisions of Exhibit "C," less the estimated cost
30 of salvaging and the estimated cost of plugging and abandoning and restoring the surface; provided, however, that in the event
31 the estimated plugging and abandoning and surface restoration costs and the estimated cost of salvaging are higher than the
32 value of the well's salvable material and equipment, each of the abandoning parties shall tender to the parties continuing
33 operations their proportionate shares of the estimated excess cost. Each abandoning party shall assign to the non-abandoning
34 parties, without warranty, express or implied, as to title or as to quantity, or fitness for use of the equipment and material, all
35 of its interest in the wellbore of the well and related equipment, together with its interest in the Leasehold insofar and only
36 insofar as such Leasehold covers the right to obtain production from that wellbore in the Zone then open to production. If the
37 interest of the abandoning party is or includes an Oil and Gas Interest, such party shall execute and deliver to the non-
38 abandoning party or parties an oil and gas lease, limited to the wellbore and the Zone then open to production, for a term of

39 one (1) year and so long thereafter as Oil and/or Gas is produced from the Zone covered thereby, such lease to be on the form

40 attached as Exhibit "B." The assignments or leases so limited shall encompass the Drilling Unit upon which the well is located.

41 The payments by, and the assignments or leases to, the assignees shall be in a ratio based upon the relationship of their

42 respective percentage of participation in the Contract Area to the aggregate of the percentages of participation in the Contract

43 Area of all assignees. There shall be no readjustment of interests in the remaining portions of the Contract Area.

44 Thereafter, abandoning parties shall have no further responsibility, liability, or interest in the operation of or production

45 from the well in the Zone then open other than the royalties retained in any lease made under the terms of this Article. Upon

46 request, Operator shall continue to operate the assigned well for the account of the non-abandoning parties at the rates and

47 charges contemplated by this agreement, plus any additional cost and charges which may arise as the result of the separate

48 ownership of the assigned well. Upon proposed abandonment of the producing Zone assigned or leased, the assignor or lessor

49 shall then have the option to repurchase its prior interest in the well (using the same valuation formula) and participate in

50 further operations therein subject to the provisions hereof.

51 3. Abandonment of Non-Consent Operations: The provisions of Article VI.E.1. or VI.E.2 above shall be applicable as

52 between Consenting Parties in the event of the proposed abandonment of any well excepted from said Articles; provided,

53 however, no well shall be permanently plugged and abandoned unless and until all parties having the right to conduct further

54 operations therein have been notified of the proposed abandonment and afforded the opportunity to elect to take over the well

55 in accordance with the provisions of this Article VI.E.; and provided further, that Non-Consenting Parties who own an interest

56 in a portion of the well shall pay their proportionate shares of abandonment and surface restoration costs for such well as

57 provided in Article VI.B.2.(b).

58 F. Termination of Operations:

59 Upon the commencement of an operation for the drilling, Reworking, Sidetracking, Plugging Back, Deepening, testing,

60 Completion or plugging of a well, including but not limited to the Initial Well, such operator shall not be terminated without

61 consent of parties bearing _____ % of the costs of such operation; provided, however, that in the event granite or other

62 practically impenetrable substance or condition in the hole is encountered which renders further operations impractical,

63 Operator may discontinue operations and give notice of such condition in the manner provided in Article VI.B.1, and the

64 provisions of Article VI.B. or VI.E. shall thereafter apply to such operation, as appropriate.

65 G. Taking Production in Kind:

66 ☐ Option No. 1: Gas Balancing Agreement Attached

67 Each party shall take in kind or separately dispose of its proportionate share of all Oil and Gas produced from the

68 Contract Area, exclusive of production which may be used in development and producing operations and in preparing and

69 treating Oil and Gas for marketing purposes and production unavoidably lost. Any extra expenditure incurred in the taking

70 in kind or separate disposition by any party of its proportionate share of the production shall be borne by such party. Any

71 party taking its share of production in kind shall be required to pay for only its proportionate share of such part of

72 Operator's surface facilities which it uses.

73 Each party shall execute such division orders and contracts as may be necessary for the sale of its interest in

74 production from the Contract Area, and, except as provided in Article VII.B, shall be entitled to receive payment

- 10 -

A.A.P.L. FORM 610 - MODEL FORM OPERATING AGREEMENT - 1989

1 directly from the purchaser thereof for its share of all production.

2 If any party fails to make the arrangements necessary to take in kind or separately dispose of its proportionate
3 share of the Oil produced from the Contract Area, Operator shall have the right, subject to the revocation at will by
4 the party owning it, but not the obligation, to purchase such Oil or sell it to others at any time and from time to
5 time, for the account of the non-taking party. Any such purchase or sale by Operator may be terminated by
6 Operator upon at least ten (10) days written notice to the owner of said production and shall be subject always to
7 the right of the owner of the production upon at least ten (10) days written notice to Operator to exercise at any
8 time its right to take in kind, or separately dispose of, its share of all Oil not previously delivered to a purchaser.
9 Any purchase or sale by Operator of any other party's share of Oil shall be only for such reasonable periods of time
10 as are consistent with the minimum needs of the industry under the particular circumstances, but in no event for a
11 period in excess of one (1) year.

12 Any such sale by Operator shall be in a manner commercially reasonable under the circumstances but Operator
13 shall have no duty to share any existing market or to obtain a price equal to that received under any existing
14 market. The sale or delivery by Operator of a non-taking party's share of Oil under the terms of any existing
15 contract of Operator shall not give the non-taking party any interest in or make the non-taking party a party to said
16 contract. No purchase shall be made by Operator without first giving the non-taking party at least ten (10) days
17 written notice of such intended purchase and the price to be paid or the pricing basis to be used.

18 All parties shall give timely written notice to Operator of their Gas marketing arrangements for the following
19 month, excluding price, and shall notify Operator immediately in the event of a change in such arrangements.
20 Operator shall maintain records of all marketing arrangements, and of volumes actually sold or transported, which
21 records shall be made available to Non-Operators upon reasonable request.

22 In the event one or more parties' separate disposition of its share of the Gas causes split-stream deliveries to separate
23 pipelines and/or deliveries which on a day-to-day basis for any reason are not exactly equal to a party's respective proportion-
24 ate share of total Gas sales to be allocated to it, the balancing or accounting between the parties shall be in accordance with
25 any Gas balancing agreement between the parties hereto, whether such an agreement is attached as Exhibit "E" or is a
26 separate agreement. Operator shall give notice to all parties of the first sales of Gas from any well under this agreement.

27 □ Option No. 2: No Gas Balancing Agreement:

28 Each party shall take in kind or separately dispose of its proportionate share of all Oil and Gas produced from
29 the Contract Area, exclusive of production which may be used in development and producing operations and in
30 preparing and treating Oil and Gas for marketing purposes and production unavoidably lost. Any extra expenditure
31 incurred in the taking in kind or separate disposition by any party of its proportionate share of the production shall
32 be borne by such party. Any party taking its share of production in kind shall be required to pay for only its
33 proportionate share of such part of Operator's surface facilities which it uses.

34 Each party shall execute such division orders and contracts as may be necessary for the sale of its interest in
35 production from the Contract Area, and, except as provided in Article VII.B, shall be entitled to receive payment
36 directly from the purchaser thereof for its share of all production.

37 If any party fails to make the arrangements necessary to take in kind or separately dispose of its proportionate

share of the Oil and/or Gas produced from the Contract Area, Operator shall have the right, subject to the revocation at will by the party owning it, but not the obligation, to purchase such Oil and/or Gas or sell it to others at any time and from time to time, for the account of the non-taking party. Any such purchase or sale by Operator may be terminated by Operator upon at least ten (10) days written notice to the owner of said production and shall be subject always to the right of the owner of the production upon at least ten (10) days written notice to Operator to exercise its right to take in kind, or separately dispose of, its share of all Oil and/or Gas not previously delivered to a purchaser; provided, however, that the effective date of any such revocation may be deferred at Operator's election for a period not to exceed ninety (90) days if Operator has committed such production to a purchase contract having a term extending beyond such ten (10) -day period. Any purchase or sale by Operator of any other party's share of Oil and/or Gas shall be only for such reasonable periods of time as are consistent with the minimum needs of the industry under the particular circumstances, but in no event for a period in excess of one (1) year.

Any such sale by Operator shall be in a manner commercially reasonable under the circumstances, but Operator shall have no duty to share any existing market or transportation arrangement or to obtain a price or transportation fee equal to that received under any existing market or transportation arrangement. The sale or delivery by Operator of a non-taking party's share of production under the terms of any existing contract of Operator shall not give the non-taking party any interest in or make the non-taking party a party to said contract. No purchase of Oil and Gas and no sale of Gas shall be made by Operator without first giving the non-taking party ten days written notice of such intended purchase or sale and the price to be paid or the pricing basis to be used. Operator shall give notice to all parties of the first sale of Gas from any well under this Agreement.

All parties shall give timely written notice to Operator of their Gas marketing arrangements for the following month, excluding price, and shall notify Operator immediately in the event of a change in such arrangements. Operator shall maintain records of all marketing arrangements, and of volumes actually sold or transported, which records shall be made available to Non-Operators upon reasonable request.

ARTICLE VII.
EXPENDITURES AND LIABILITY OF PARTIES

A. Liability of Parties:

The liability of the parties shall be several, not joint or collective. Each party shall be responsible only for its obligations, and shall be liable only for its proportionate share of the costs of developing and operating the Contract Area. Accordingly, the liens granted among the parties in Article VII.B. are given to secure only the debts of each severally, and no party shall have any liability to third parties hereunder to satisfy the default of any other party in the payment of any expense or obligation hereunder. It is not the intention of the parties to create, nor shall this agreement be construed as creating, a mining or other partnership, joint venture, agency relationship or association, or to render the parties liable as partners, co-venturers, or principals. In their relations with each other under this agreement, the parties shall not be considered fiduciaries or to have established a confidential relationship but rather shall be free to act on an arm's-length basis in accordance with their own respective self-interest, subject, however, to the obligation of the parties to act in good faith in their dealings with each other with respect to activities hereunder.

38
39
40
41
42
43
44
45
46
47
48
49
50
51
52
53
54
55
56
57
58
59
60
61
62
63
64
65
66
67
68
69
70
71
72
73
74

- 11 -

A.A.P.L. FORM 610 - MODEL FORM OPERATING AGREEMENT - 1989

B. Liens and Security Interests:

1
2   Each party grants to the other parties hereto a lien upon any interest it now owns or hereafter acquires in Oil and Gas
3 Leases and Oil and Gas Interests in the Contract Area, and a security interest and/or purchase money security interest in any
4 interest it now owns or hereafter acquires in the personal property and fixtures on or used or obtained for use in connection
5 therewith, to secure performance of all of its obligations under this agreement including but not limited to payment of expense,
6 interest and fees, the proper disbursement of all monies paid hereunder, the assignment or relinquishment of interest in Oil
7 and Gas Leases as required hereunder, and the proper performance of operations hereunder. Such lien and security interest
8 granted by each party hereto shall include such party's leasehold interests, working interests, operating rights, and royalty and
9 overriding royalty interests in the Contract Area now owned or hereafter acquired and in lands pooled or unitized therewith or
10 otherwise becoming subject to this agreement, the Oil and Gas when extracted therefrom and equipment situated thereon or
11 used or obtained for use in connection therewith (including, without limitation, all wells, tools, and tubular goods), and accounts
12 (including, without limitation, accounts arising from gas imbalances or from the sale of Oil and/or Gas at the wellhead),
13 contract rights, inventory and general intangibles relating thereto or arising therefrom, and all proceeds and products of the
14 foregoing.
15   To perfect the lien and security agreement provided herein, each party hereto shall execute and acknowledge the recording
16 supplement and/or any financing statement prepared and submitted by any party hereto in conjunction herewith or at any time
17 following execution hereof, and Operator is authorized to file this agreement or the recording supplement executed herewith as
18 a lien or mortgage in the applicable real estate records and as a financing statement with the proper officer under the Uniform
19 Commercial Code in the state in which the Contract Area is situated and such other states as Operator shall deem appropriate
20 to perfect the security interest granted hereunder. Any party may file this agreement, the recording supplement executed
21 herewith, or such other documents as it deems necessary as a lien or mortgage in the applicable real estate records and/or a
22 financing statement with the proper officer under the Uniform Commercial Code.
23   Each party represents and warrants to the other parties hereto that the lien and security interest granted by such party to
24 the other parties shall be a first and prior lien, and each party hereby agrees to maintain the priority of said lien and security
25 interest against all persons acquiring an interest in Oil and Gas Leases and Interests covered by this agreement by, through or
26 under such party. All parties acquiring an interest in Oil and Gas Leases and Oil and Gas Interests covered by this agreement,
27 whether by assignment, merger, mortgage, operation of law, or otherwise, shall be deemed to have taken subject
28 to the lien and security interest granted by this Article VII.B. as to all obligations attributable to such interest hereunder
29 whether or not such obligations arise before or after such interest is acquired.
30   To the extent that parties have a security interest under the Uniform Commercial Code of the state in which the
31 Contract Area is situated, they shall be entitled to exercise the rights and remedies of a secured party under the Code.
32 The bringing of a suit and the obtaining of judgment by a party for the secured indebtedness shall not be deemed an
33 election of remedies or otherwise affect the lien rights or security interest as security for the payment thereof. In
34 addition, upon default by any party in the payment of its share of expenses, interests or fees, or upon the improper use
35 of funds by the Operator, the other parties shall have the right, without prejudice to other rights or remedies, to collect
36 from the purchaser the proceeds from the sale of such defaulting party's share of Oil and Gas until the amount owed by
37 such party, plus interest as provided in "Exhibit C," has been received, and shall have the right to offset the amount
38 owed against the proceeds from the sale of such defaulting party's share of Oil and Gas. All purchasers of production

39   may rely on a notification of default from the non-defaulting party or parties stating the amount due as a result of the
40   default, and all parties waive any recourse available against purchasers for releasing production proceeds as provided in
41   this paragraph.
42   If any party fails to pay its share of cost within one hundred twenty (120) days after rendition of a statement therefor by
43   Operator, the non-defaulting parties, including Operator, shall, upon request by Operator, pay the unpaid amount in the
44   proportion that the interest of each such party bears to the interest of all such parties. The amount paid by each party so
45   paying its share of the unpaid amount shall be secured by the liens and security rights described in Article VII.B, and each
46   paying party may independently pursue any remedy available hereunder or otherwise.
47   If any party does not perform all of its obligations hereunder, and the failure to perform subjects such party to foreclosure
48   or execution proceedings pursuant to the provisions of this agreement, to the extent allowed by governing law, the defaulting
49   party waives any available right of redemption from and after the date of judgment, any required valuation or appraisement
50   of the mortgaged or secured property prior to sale, any available right to stay execution or to require a marshalling of assets
51   and any required bond in the event a receiver is appointed. In addition, to the extent permitted by applicable law, each party
52   hereby grants to the other parties a power of sale as to any property that is subject to the lien and security rights granted
53   hereunder, such power to be exercised in the manner provided by applicable law or otherwise in a commercially reasonable
54   manner and upon reasonable notice.
55   Each party agrees that the other parties shall be entitled to utilize the provisions of Oil and Gas lien law or other lien
56   law of any state in which the Contract Area is situated to enforce the obligations of each party hereunder. Without limiting
57   the generality of the foregoing, to the extent permitted by applicable law, Non-Operators agree that Operator may invoke or
58   utilize the mechanics' or materialmen's lien law of the state in which the Contract Area is situated in order to secure the
59   payment to Operator of any sum due hereunder for services performed or materials supplied by Operator.
60   **C. Advances:**
61   Operator, at its election, shall have the right from time to time to demand and receive from one or more of the other
62   parties payment in advance of their respective shares of the estimated amount of the expense to be incurred in operations
63   hereunder during the next succeeding month, which right may be exercised only by submission to each such party of an
64   itemized statement of such estimated expense, together with an invoice for its share thereof. Each such statement and invoice
65   for the payment in advance of estimated expense shall be submitted on or before the 20th day of the next preceding month.
66   Each party shall pay to Operator its proportionate share of such estimate within fifteen (15) days after such estimate and
67   invoice is received. If any party fails to pay its share of said estimate within said time, the amount due shall bear interest as
68   provided in Exhibit "C" until paid. Proper adjustment shall be made monthly between advances and actual expense to the end
69   that each party shall bear and pay its proportionate share of actual expenses incurred, and no more.
70   **D. Defaults and Remedies:**
71   If any party fails to discharge any financial obligation under this agreement, including without limitation the failure to
72   make any advance under the preceding Article VII.C. or any other provision of this agreement, within the period required for
73   such payment hereunder, then in addition to the remedies provided in Article VII.B. or elsewhere in this agreement, the
74   remedies specified below shall be applicable. For purposes of this Article VII.D, all notices and elections shall be delivered

- 12 -

227

A.A.P.L. FORM 610 - MODEL FORM OPERATING AGREEMENT - 1989

1 only by Operator, except that Operator shall deliver any such notice and election requested by a non-defaulting Non-Operator,
2 and when Operator is the party in default, the applicable notices and elections can be delivered by any Non-Operator.
3 Election of any one or more of the following remedies shall not preclude the subsequent use of any other remedy specified
4 below or otherwise available to a non-defaulting party.

5   1. Suspension of Rights: Any party may deliver to the party in default a Notice of Default, which shall specify the default,
6 specify the action to be taken to cure the default, and specify that failure to take such action will result in the exercise of one
7 or more of the remedies provided in this Article. If the default is not cured within thirty (30) days of the delivery of such
8 Notice of Default, all of the rights of the defaulting party granted by this agreement may upon notice be suspended until the
9 default is cured, without prejudice to the right of the non-defaulting party or parties to continue to enforce the obligations of
10 the defaulting party previously accrued or thereafter accruing under this agreement. If Operator is the party in default, the
11 Non-Operators shall have in addition the right, by vote of Non-Operators owning a majority in interest in the Contract Area
12 after excluding the voting interest of Operator, to appoint a new Operator effective immediately. The rights of a defaulting
13 party that may be suspended hereunder at the election of the non-defaulting parties shall include, without limitation, the right
14 to receive information as to any operation conducted hereunder during the period of such default, the right to elect to
15 participate in an operation proposed under Article VI.B. of this agreement, the right to participate in an operation being
16 conducted under this agreement even if the party has previously elected to participate in such operation, and the right to
17 receive proceeds of production from any well subject to this agreement.

18   2. Suit for Damages: Non-defaulting parties or Operator for the benefit of non-defaulting parties may sue (at joint
19 account expense) to collect the amounts in default, plus interest accruing on the amounts recovered from the date of default
20 until the date of collection at the rate specified in Exhibit "C" attached hereto. Nothing herein shall prevent any party from
21 suing any defaulting party to collect consequential damages accruing to such party as a result of the default.

22   3. Deemed Non-Consent: The non-defaulting party may deliver a written Notice of Non-Consent Election to the
23 defaulting party at any time after the expiration of the thirty-day cure period following delivery of the Notice of Default, in
24 which event if the billing is for the drilling of a new well or the Plugging Back, Sidetracking, Reworking or Deepening of a
25 well which is to be or has been plugged as a dry hole, or for the Completion or Recompletion of any well, the defaulting
26 party will be conclusively deemed to have elected not to participate in the operation and to be a Non-Consenting Party with
27 respect thereto under Article VI.B. or VI.C., as the case may be, to the extent of the costs unpaid by such party,
28 notwithstanding any election to participate theretofore made. If election is made to proceed under this provision, then the
29 non-defaulting parties may not elect to sue for the unpaid amount pursuant to Article VII.D.2.

30   Until the delivery of such Notice of Non-Consent Election to the defaulting party, such party shall have the right to cure
31 its default by paying its unpaid share of costs plus interest at the rate set forth in Exhibit "C," provided, however, such
32 payment shall not prejudice the rights of the non-defaulting parties to pursue remedies for damages incurred by the non-
33 defaulting parties as a result of the default. Any interest relinquished pursuant to this Article VII.D.3. shall be offered to the
34 non-defaulting parties in proportion to their interests, and the non-defaulting parties electing to participate in the ownership
35 of such interest shall be required to contribute their shares of the defaulted amount upon their election to participate therein.

36   4. Advance Payment: If a default is not cured within thirty (30) days of the delivery of a Notice of Default, Operator, or
37 Non-Operators if Operator is the defaulting party, may thereafter require advance payment from the defaulting
38 party of such defaulting party's anticipated share of any item of expense for which Operator, or Non-Operators, as the case may

39   be, would be entitled to reimbursement under any provision of this agreement, whether or not such expense was the subject of

40   the previous default. Such right includes, but is not limited to, the right to require advance payment for the estimated costs of

41   drilling a well or Completion of a well as to which an election to participate in drilling or Completion has been made. If the

42   defaulting party fails to pay the required advance payment, the non-defaulting parties may pursue any of the remedies provided

43   in this Article VII.D. or any other default remedy provided elsewhere in this agreement. Any excess of funds advanced remaining

44   when the operation is completed and all costs have been paid shall be promptly returned to the advancing party.

45   5. Costs and Attorneys' Fees. In the event any party is required to bring legal proceedings to enforce any financial

46   obligation of a party hereunder, the prevailing party in such action shall be entitled to recover all court costs, costs of

47   collection, and a reasonable attorney's fee, which the lien provided for herein shall also secure.

48   **E. Rentals, Shut-in Well Payments and Minimum Royalties:**

49   Rentals, shut-in well payments and minimum royalties which may be required under the terms of any lease shall be paid

50   by the party or parties who subjected such lease to this agreement at its or their expense. In the event two or more parties

51   own and have contributed interests in the same lease to this agreement, such parties may designate one of such parties to

52   make said payments for and on behalf of all such parties. Any party may request, and shall be entitled to receive, proper

53   evidence of all such payments. In the event of failure to make proper payment of any rental, shut-in well payment or

54   minimum royalty through mistake or oversight where such payment is required to continue the lease in force, any loss which

55   results from such non-payment shall be borne in accordance with the provisions of Article IV.B.2.

56   Operator shall notify Non-Operators of the anticipated completion of a shut-in well, or the shutting in or return to

57   production of a producing well, at least five (5) days (excluding Saturday, Sunday and legal holidays) prior to taking such

58   action, or at the earliest opportunity permitted by circumstances, but assumes no liability for failure to do so. In the event of

59   failure by Operator to so notify Non-Operators, the loss of any lease contributed hereto by Non-Operators for failure to make

60   timely payments of any shut-in well payment shall be borne jointly by the parties hereto under the provisions of Article

61   IV.B.3.

62   **F. Taxes:**

63   Beginning with the first calendar year after the effective date hereof, Operator shall render for ad valorem taxation all

64   property subject to this agreement which by law should be rendered for such taxes, and it shall pay all such taxes assessed

65   thereon before they become delinquent. Prior to the rendition date, each Non-Operator shall furnish Operator information as

66   to burdens (to include, but not be limited to, royalties, overriding royalties and production payments) on Leases and Oil and

67   Gas Interests contributed by such Non-Operator. If the assessed valuation of any Lease is reduced by reason of its being

68   subject to outstanding excess royalties, overriding royalties or production payments, the reduction in ad valorem taxes

69   resulting therefrom shall inure to the benefit of the owner or owners of such Lease, and Operator shall adjust the charge to

70   such owner or owners so as to reflect the benefit of such reduction. If the ad valorem taxes are based in whole or in part

71   upon separate valuations of each party's working interest, then notwithstanding anything to the contrary herein, charges to

72   the joint account shall be made and paid by the parties hereto in accordance with the tax value generated by each party's

73   working interest. Operator shall bill the other parties for their proportionate shares of all tax payments in the manner

74   provided in Exhibit "C."

- 13 -

A.A.P.L. FORM 610 - MODEL FORM OPERATING AGREEMENT - 1989

1.   If Operator considers any tax assessment improper, Operator may, at its discretion, protest within the time and manner
2. prescribed by law, and prosecute the protest to a final determination, unless all parties agree to abandon the protest prior to final
3. determination. During the pendency of administrative or judicial proceedings, Operator may elect to pay, under protest, all such taxes
4. and any interest and penalty. When any such protested assessment shall have been finally determined, Operator shall pay the tax for
5. the joint account, together with any interest and penalty accrued, and the total cost shall then be assessed against the parties, and be
6. paid by them, as provided in Exhibit "C."
7.   Each party shall pay or cause to be paid all production, severance, excise, gathering and other taxes imposed upon or with respect
8. to the production or handling of such party's share of Oil and Gas produced under the terms of this agreement.

ARTICLE VIII.

ACQUISITION, MAINTENANCE OR TRANSFER OF INTEREST

A. Surrender of Leases:

12.   The Leases covered by this agreement, insofar as they embrace acreage in the Contract Area, shall not be surrendered in whole
13. or in part unless all parties consent thereto.
14.   However, should any party desire to surrender its interest in any Lease or in any portion thereof, such party shall give written
15. notice of the proposed surrender to all parties, and the parties to whom such notice is delivered shall have thirty (30) days after
16. delivery of the notice within which to notify the party proposing the surrender whether they elect to consent thereto. Failure of a
17. party to whom such notice is delivered to reply within said 30-day period shall constitute a consent to the surrender of the Leases
18. described in the notice. If all parties do not agree or consent thereto, the party desiring to surrender shall assign, without express or
19. implied warranty of title, all of its interest in such Lease, or portion thereof, and any well, material and equipment which may be
20. located thereon and any rights in production thereafter secured, to the parties not consenting to such surrender. If the interest of the
21. assigning party is or includes an Oil and Gas Interest, the assigning party shall execute and deliver to the party or parties not
22. consenting to such surrender an oil and gas lease covering such Oil and Gas Interest for a term of one (1) year and so long
23. thereafter as Oil and/or Gas is produced from the land covered thereby, such lease to be on the form attached hereto as Exhibit "B."
24.   Upon such assignment or lease, the assigning party shall be relieved from all obligations thereafter accruing, but not theretofore
25. accrued, with respect to the interest assigned or leased and the operation of any well attributable thereto, and the assigning party
26. shall have no further interest in the assigned or leased premises and its equipment and production other than the royalties retained
27. in any lease made under the terms of this Article. The party assignee or lessee shall pay to the party assignor or lessor the
28. reasonable salvage value of the latter's interest in any well's salvable materials and equipment attributable to the assigned or leased
29. acreage. The value of all salvable materials and equipment shall be determined in accordance with the provisions of Exhibit "C," less
30. the estimated cost of salvaging and the estimated cost of plugging and abandoning and restoring the surface. If such value is less
31. than such costs, then the party assignor or lessor shall pay to the party assignee or lessee the amount of such deficit. If the
32. assignment or lease is in favor of more than one party, the interest shall be shared by such parties in the proportions that the
33. interest of each bears to the total interest of all such parties. If the interest of the parties to whom the assignment is to be made
34. varies according to depth, then the interest assigned shall similarly reflect such variances.
35.   Any assignment, lease or surrender made under this provision shall not reduce or change the assignor's, lessor's or surrendering
36. party's interest as it was immediately before the assignment, lease or surrender in the balance of the Contract Area; and the acreage
37. assigned, leased or surrendered, and subsequent operations thereon, shall not thereafter be subject to the terms and provisions of this
38. agreement but shall be deemed subject to an Operating Agreement in the form of this agreement.

B. Renewal or Extension of Leases:

If any party secures a renewal or replacement of an Oil and Gas Lease or Interest subject to this agreement, then all other parties shall be notified promptly upon such acquisition or, in the case of a replacement Lease taken before expiration of an existing Lease, promptly upon expiration of the existing Lease. The parties notified shall have the right for a period of thirty (30) days following delivery of such notice in which to elect to participate in the ownership of the renewal or replacement Lease, insofar as such Lease affects lands within the Contract Area, by paying to the party who acquired it their proportionate shares of the acquisition cost allocated to that part of such Lease within the Contract Area, which shall be in proportion to the interests held at that time by the parties in the Contract Area. Each party who participates in the purchase of a renewal or replacement Lease shall be given an assignment of its proportionate interest therein by the acquiring party.

If some, but less than all, of the parties elect to participate in the purchase of a renewal or replacement Lease, it shall be owned by the parties who elect to participate therein, in a ratio based upon the relationship of their respective percentage of participation in the Contract Area to the aggregate of the percentages of participation in the Contract Area of all parties participating in the purchase of such renewal or replacement Lease. The acquisition of a renewal or replacement Lease by any or all of the parties hereto shall not cause a readjustment of the interests of the parties stated in Exhibit "A," but any renewal or replacement Lease in which less than all parties elect to participate shall not be subject to this agreement but shall be deemed subject to a separate Operating Agreement in the form of this agreement.

If the interests of the parties in the Contract Area vary according to depth, then their right to participate proportionately in renewal or replacement Leases and their right to receive an assignment of interest shall also reflect such depth variances.

The provisions of this Article shall apply to renewal or replacement Leases whether they are for the entire interest covered by the expiring Lease or cover only a portion of its area or an interest therein. Any renewal or replacement Lease taken before the expiration of its predecessor Lease, or taken or contracted for or becoming effective within six (6) months after the expiration of the existing Lease, shall be subject to this provision so long as this agreement is in effect at the time of such acquisition or at the time the renewal or replacement Lease becomes effective; but any Lease taken or contracted for more than six (6) months after the expiration of an existing Lease shall not be deemed a renewal or replacement Lease and shall not be subject to the provisions of this agreement.

The provisions in this Article shall also be applicable to extensions of Oil and Gas Leases.

C. Acreage or Cash Contributions:

While this agreement is in force, if any party contracts for a contribution of cash towards the drilling of a well or any other operation on the Contract Area, such contribution shall be paid to the party who conducted the drilling or other operation and shall be applied by it against the cost of such drilling or other operation. If the contribution be in the form of acreage, the party to whom the contribution is made shall promptly tender an assignment of the acreage, without warranty of title, to the Drilling Parties in the proportions said Drilling Parties shared the cost of drilling the well. Such acreage shall become a separate Contract Area and, to the extent possible, be governed by provisions identical to this agreement. Each party shall promptly notify all other parties of any acreage or cash contributions it may obtain in support of any well or any other operation on the Contract Area. The above provisions shall also be applicable to optional rights to earn acreage outside the Contract Area which are in support of well drilled inside the Contract Area.

A.A.P.L. FORM 610 - MODEL FORM OPERATING AGREEMENT - 1989

1 If any party contracts for any consideration relating to disposition of such party's share of substances produced hereunder,
2 such consideration shall not be deemed a contribution as contemplated in this Article VIII.C.

3 **D. Assignment; Maintenance of Uniform Interest:**

4 For the purpose of maintaining uniformity of ownership in the Contract Area in the Oil and Gas Leases, Oil and Gas
5 Interests, wells, equipment and production covered by this agreement no party shall sell, encumber, transfer or make other
6 disposition of its interest in the Oil and Gas Leases and Oil and Gas Interests embraced within the Contract Area or in wells,
7 equipment and production unless such disposition covers either:

8 1. the entire interest of the party in all Oil and Gas Leases, Oil and Gas Interests, wells, equipment and production; or
9 2. an equal undivided percent of the party's present interest in all Oil and Gas Leases, Oil and Gas Interests, wells,
10 equipment and production in the Contract Area.

11 Every sale, encumbrance, transfer or other disposition made by any party shall be made expressly subject to this agreement
12 and shall be made without prejudice to the right of the other parties, and any transferee of an ownership interest in any Oil and
13 Gas Lease or Interest shall be deemed a party to this agreement as to the interest conveyed from and after the effective date of
14 the transfer of ownership; provided, however, that the other parties shall not be required to recognize any such sale,
15 encumbrance, transfer or other disposition for any purpose hereunder until thirty (30) days after they have received a copy of the
16 instrument of transfer or other satisfactory evidence thereof in writing from the transferor or transferee. No assignment or other
17 disposition of interest by a party shall relieve such party of obligations previously incurred by such party hereunder with respect
18 to the interest transferred, including without limitation the obligation of a party to pay all costs attributable to an operation
19 conducted hereunder in which such party has agreed to participate prior to making such assignment, and the lien and security
20 interest granted by Article VII.B. shall continue to burden the interest transferred to secure payment of any such obligations.

21 If, at any time the interest of any party is divided among and owned by four or more co-owners, Operator, at its discretion,
22 may require such co-owners to appoint a single trustee or agent with full authority to receive notices, approve expenditures,
23 receive billings for and approve and pay such party's share of the joint expenses, and to deal generally with, and with power to
24 bind, the co-owners of such party's interest within the scope of the operations embraced in this agreement; however, all such co-
25 owners shall have the right to enter into and execute all contracts or agreements for the disposition of their respective shares of
26 the Oil and Gas produced from the Contract Area and they shall have the right to receive, separately, payment of the sale
27 proceeds thereof.

28 **E. Waiver of Rights to Partition:**

29 If permitted by the laws of the state or states in which the property covered hereby is located, each party hereto owning an
30 undivided interest in the Contract Area waives any and all rights it may have to partition and have set aside to it in severalty its
31 undivided interest therein.

32 **F. Preferential Right to Purchase:**

33 □ (Optional; Check if applicable.)

34 Should any party desire to sell all or any part of its interests under this agreement, or its rights and interests in the Contract
35 Area, it shall promptly give written notice to the other parties, with full information concerning its proposed disposition, which
36 shall include the name and address of the prospective transferee (who must be ready, willing and able to purchase), the purchase
37 price, a legal description sufficient to identify the property, and all other terms of the offer. The other parties shall then have an
38 optional prior right, for a period of ten (10) days after the notice is delivered, to purchase for the stated consideration on the

same terms and conditions the interest which the other party proposes to sell; and, if this optional right is exercised, the purchasing parties shall share the purchased interest in the proportions that the interest of each bears to the total interest of all purchasing parties. However, there shall be no preferential right to purchase in those cases where any party wishes to mortgage its interests, or to transfer title to its interests to its mortgagee in lieu of or pursuant to foreclosure of a mortgage of its interests, or to dispose of its interests by merger, reorganization, consolidation, or by sale of all or substantially all of its Oil and Gas assets to any party, or by transfer of its interests to a subsidiary or parent company or to a subsidiary of a parent company, or to any company in which such party owns a majority of the stock.

ARTICLE IX.
INTERNAL REVENUE CODE ELECTION

If, for federal income tax purposes, this agreement and the operations hereunder are regarded as a partnership, and if the parties have not otherwise agreed to form a tax partnership pursuant to Exhibit "G" or other agreement between them, each party thereby affected elects to be excluded from the application of all of the provisions of Subchapter "K," Chapter 1, Subtitle "A," of the Internal Revenue Code of 1986, as amended ("Code"), as permitted and authorized by Section 761 of the Code and the regulations promulgated thereunder. Operator is authorized and directed to execute on behalf of each party hereby affected such evidence of this election as may be required by the Secretary of the Treasury of the United States or the Federal Internal Revenue Service, including specifically, but not by way of limitation, all of the returns, statements, and the data required by Treasury Regulations §1.761. Should there be any requirement that each party hereby affected give further evidence of this election, each such party shall execute such documents and furnish such other evidence as may be required by the Federal Internal Revenue Service or as may be necessary to evidence this election. No such party shall give any notices or take any other action inconsistent with the election made hereby. If any present or future income tax laws of the state or states in which the Contract Area is located or any future tax laws of the United States contain provisions similar to those in Subchapter "K," Chapter 1, Subtitle "A," of the Code, under which an election similar to that provided by Section 761 of the Code is permitted, each party hereby affected shall make such election as may be permitted or required by such laws. In making the foregoing election, each such party states that the income derived by such party from operations hereunder can be adequately determined without the computation of partnership taxable income.

ARTICLE X.
CLAIMS AND LAWSUITS

Operator may settle any single uninsured third party damage claim or suit arising from operations hereunder if the expenditure does not exceed _____ Dollars ($ _____) and if the payment is in complete settlement of such claim or suit. If the amount required for settlement exceeds the above amount, the parties hereto shall assume and take over the further handling of the claim or suit, unless such authority is delegated to Operator. All costs and expenses of handling, settling, or otherwise discharging such claim or suit shall be at the joint expense of the parties participating in the operation from which the claim or suit arises. If a claim is made against any party or if any party is sued on account of any matter arising from operations hereunder over which such individual has no control because of the rights given Operator by this agreement, such party shall immediately notify all other parties, and the claim or suite shall be treated as any other claim or suit involving operations hereunder.

- 15 -

A.A.P.L. FORM 610 - MODEL FORM OPERATING AGREEMENT - 1989

ARTICLE XI.
FORCE MAJEURE

1
2   If any party is rendered unable, wholly or in part, by force majeure to carry out its obligations under this agreement, other
3   than the obligation to indemnify or make money payments or furnish security, that party shall give to all other parties
4   prompt written notice of the force majeure with reasonably full particulars concerning it; thereupon, the obligations of the
5   party giving the notice, so far as they are affected by the force majeure, shall be suspended during, but no longer than, the
6   continuance of the force majeure. The term "force majeure," as here employed, shall mean an act of God, strike, lockout, or
7   other industrial disturbance, act of the public enemy, war, blockade, public riot, lightning, fire, storm, flood or other act of
8   nature, explosion, governmental action, governmental delay, restraint or inaction, unavailability of equipment, and any other
9   cause, whether of the kind specifically enumerated above or otherwise, which is not reasonably within the control of the party
10 claiming suspension.
11
12 The affected party shall use all reasonable diligence to remove the force majeure situation as quickly as practicable. The
13 requirement that any force majeure shall be remedied with all reasonable dispatch shall not require the settlement of strikes,
14 lockouts, or other labor difficulty by the party involved, contrary to its wishes; how all such difficulties shall be handled shall
15 be entirely within the discretion of the party concerned.
16

ARTICLE XII.
NOTICES

17
18 All notices authorized or required between the parties by any of the provisions of this agreement, unless otherwise
19 specifically provided, shall be in writing and delivered in person or by United States mail, courier service, telegram, telex,
20 telecopier or any other form of facsimile, postage or charges prepaid, and addressed to such parties at the addresses listed on
21 Exhibit "A." All telephone or oral notices permitted by this agreement shall be confirmed immediately thereafter by written
22 notice. The originating notice given under any provision hereof shall be deemed delivered only when received by the party to
23 whom such notice is directed, and the time for such party to deliver any notice in response thereto shall run from the date
24 the originating notice is received. "Receipt" for purposes of this agreement with respect to written notice delivered hereunder
25 shall be actual delivery of the notice to the address of the party to be notified specified in accordance with this agreement, or
26 to the telecopy, facsimile or telex machine of such party. The second or any responsive notice shall be deemed delivered when
27 deposited in the United States mail or at the office of the courier or telegraph service, or upon transmittal by telex, telecopy
28 or facsimile, or when personally delivered to the party to be notified, provided, that when response is required within 24 or
29 48 hours, such response shall be given orally or by telephone, telex, telecopy or other facsimile within such period. Each party
30 shall have the right to change its address at any time, and from time to time, by giving written notice thereof to all other
31 parties. If a party is not available to receive notice orally or by telephone when a party attempts to deliver a notice required
32 to be delivered within 24 or 48 hours, the notice may be delivered in writing by any other method specified herein and shall
33 be deemed delivered in the same manner provided above for any responsive notice.
34

ARTICLE XIII.
TERM OF AGREEMENT

35
36 This agreement shall remain in full force and effect as to the Oil and Gas Leases and/or Oil and Gas Interests subject
37 hereto for the period of time selected below; provided, however, no party hereto shall ever be construed as having any right, title

38 or interest in or to any Lease or Oil and Gas Interest contributed by any other party beyond the term of this agreement.

39 ☐ Option No. 1: So long as any of the Oil and Gas Leases subject to this agreement remain or are continued in
40 force as to any part of the Contract Area, whether by production, extension, renewal or otherwise.

41 ☐ Option No. 2: In the event the well described in Article VI.A., or any subsequent well drilled under any provision
42 of this agreement, results in the Completion of a well as a well capable of production of Oil and/or Gas in paying
43 quantities, this agreement shall continue in force so long as any such well is capable of production, and for an
44 additional period of _____ days thereafter; provided, however, if, prior to the expiration of such
45 additional period, one or more of the parties hereto are engaged in drilling, Reworking, Deepening, Sidetracking,
46 Plugging Back, testing or attempting to Complete or Re-complete a well or wells hereunder, this agreement shall
47 continue in force until such operations have been completed and if production results therefrom, this agreement
48 shall continue in force as provided herein. In the event the well described in Article VI.A., or any subsequent well
49 drilled hereunder, results in a dry hole, and no other well is capable of producing Oil and/or Gas from the
50 Contract Area, this agreement shall terminate unless drilling, Deepening, Sidetracking, Completing, Re-
51 completing, Plugging Back or Reworking operations are commenced within _____ days from the
52 date of abandonment of said well. "Abandonment" for such purposes shall mean either (i) a decision by all parties
53 not to conduct any further operations on the well or (ii) the elapse of 180 days from the conduct of any
54 operations on the well, whichever first occurs.

55 The termination of this agreement shall not relieve any party hereto from any expense, liability or other obligation or any
56 remedy therefor which has accrued or attached prior to the date of such termination.

57 Upon termination of this agreement and the satisfaction of all obligations hereunder, in the event a memorandum of this
58 Operating Agreement has been filed of record, Operator is authorized to file of record in all necessary recording offices a
59 notice of termination, and each party hereto agrees to execute such a notice of termination as to Operator's interest, upon
60 request of Operator, if Operator has satisfied all its financial obligations.

ARTICLE XIV.
COMPLIANCE WITH LAWS AND REGULATIONS

63 A. Laws, Regulations and Orders:

64 This agreement and all matters pertaining hereto, including but not limited to matters of performance, non-
65 performance, breach, remedies, procedures, rights, duties, and interpretation or construction, shall be governed and
66 determined by the law of the state in which the Contract Area is located. If the Contract Area is in two or more states,
67 the law of the state of _____ shall govern.

63 A. Laws, Regulations and Orders:

64 This agreement shall be subject to the applicable laws of the state in which the Contract Area is located, to the valid rules,
65 regulations, and orders of any duly constituted regulatory body of said state; and to all other applicable federal, state,
66 and local laws, ordinances, rules, regulations and orders.

67 B. Governing Law:

68 This agreement and all matters pertaining hereto, including but not limited to matters of performance, non-
69 performance, breach, remedies, procedures, rights, duties, and interpretation or construction, shall be governed and
70 determined by the law of the state in which the Contract Area is located. If the Contract Area is in two or more states,
71 the law of the state of _____ shall govern.

72 C. Regulatory Agencies:

73 Nothing herein contained shall grant, or be construed to grant, Operator the right or authority to waive or release any
74 rights, privileges, or obligations which Non-Operators may have under federal or state laws or under rules, regulations or

- 16 -

A.A.P.L. FORM 610 - MODEL FORM OPERATING AGREEMENT - 1989

1 orders promulgated under such laws in reference to oil, gas and mineral operations, including the location, operation, or
2 production of wells, on tracts offsetting or adjacent to the Contract Area.

3 With respect to the operations hereunder, Non-Operators agree to release Operator from any and all losses, damages,
4 injuries, claims and causes of action arising out of, incident to or resulting directly or indirectly from Operator's interpretation
5 or application of rules, rulings, regulations or orders of the Department of Energy or Federal Energy Regulatory Commission
6 or predecessor or successor agencies to the extent such interpretation or application was made in good faith and does not
7 constitute gross negligence. Each Non-Operator further agrees to reimburse Operator for such Non-Operator's share of
8 production or any refund, fine, levy or other governmental sanction that Operator may be required to pay as a result of such
9 an incorrect interpretation or application, together with interest and penalties thereon owing by Operator as a result of such
10 incorrect interpretation or application.

ARTICLE XV.
MISCELLANEOUS

A. Execution:

14 This agreement shall be binding upon each Non-Operator when this agreement or a counterpart thereof has been
15 executed by such Non-Operator and Operator notwithstanding that this agreement is not then or thereafter executed by all of
16 the parties to which it is tendered or which are listed on Exhibit "A" as owning an interest in the Contract Area or which
17 own, in fact, an interest in the Contract Area. Operator may, however, by written notice to all Non-Operators who have
18 become bound by this agreement as aforesaid, given at any time prior to the actual spud date of the Initial Well but in no
19 event later than five days prior to the date specified in Article VI.A. for commencement of the Initial Well, terminate this
20 agreement if Operator in its sole discretion determines that there is insufficient participation to justify commencement of
21 drilling operations. In the event of such a termination by Operator, all further obligations of the parties hereunder shall cease
22 as of such termination. In the event any Non-Operator has advanced or prepaid any share of drilling or other costs
23 hereunder, all sums so advanced shall be returned to such Non-Operator without interest. In the event Operator proceeds
24 with drilling operations for the Initial Well without the execution hereof by all persons listed on Exhibit "A" as having a
25 current working interest in such well, Operator shall indemnify Non-Operators with respect to all costs incurred for the
26 Initial Well which would have been charged to such person under this agreement if such person had executed the same and
27 Operator shall receive all revenues which would have been received by such person under this agreement if such person had
28 executed the same.

B. Successors and Assigns:

30 This agreement shall be binding upon and shall inure to the benefit of the parties hereto and their respective heirs,
31 devisees, legal representatives, successors and assigns, and the terms hereof shall be deemed to run with the Leases or
32 Interests included within the Contract Area.

C. Counterparts:

34 This instrument may be executed in any number of counterparts, each of which shall be considered an original for all
35 purposes.

D. Severability:

37 For the purposes of assuming or rejecting this agreement as an executory contract pursuant to federal bankruptcy laws,

this agreement shall not be severable, but rather must be assumed or rejected in its entirety, and the failure of any party to this agreement to comply with all of its financial obligations provided herein shall be a material default.

ARTICLE XVI.

OTHER PROVISIONS

38
39
40
41
42
43
44
45
46
47
48
49
50
51
52
53
54
55
56
57
58
59
60
61
62
63
64
65
66
67
68
69
70
71
72
73
74

- 17 -

237

A.A.P.L. FORM 610 - MODEL FORM OPERATING AGREEMENT - 1989

1   IN WITNESS WHEREOF, this agreement shall be effective as of the _____ day of _____ ,

2   19 _____ .

3   ATTEST OR WITNESS: OPERATOR

4   _____

5   _____   By _____

6
7   _____   Type or print name

8   Title _____

9   Date _____

10   Tax ID or S.S. No. _____

11

12   NON-OPERATORS

13

14   _____

15   _____   By _____

16
17   _____   Type or print name

18   Title _____

19   Date _____

Tax ID or S.S. No.

By

Type or print name

Title

Date

Tax ID or S.S. No.

By

Type or print name

Title

Date

Tax ID or S.S. No.

20
21
22
23
24
25
26
27
28
29
30
31
32
33
34
35
36
37

- 18 -

A.A.P.L. FORM 610 - MODEL FORM OPERATING AGREEMENT - 1989

ACKNOWLEDGMENTS

Note: The following forms of acknowledgment are the short forms approved by the Uniform Law on Notarial Acts. The validity and effect of these forms in any state will depend upon the statutes of that state.

Individual acknowledgment:

State of ————————)
) ss.
County of ————————)

This instrument was acknowledged before me on ———————————— by ————————————.

————————————————————

(Seal, if any)

Title (and Rank) ————————————————

My commission expires: ————————————————

Acknowledgment in representative capacity:

State of ————————)
) ss.
County of ————————)

20

This instrument was acknowledged before me on

21 _____ by _____ as

22 _____ of _____ .

23 (Seal, if any)

24 _____ Title (and Rank)

25 _____ My commission expires

26

27

28

29

30

31

32

33

34

35

36

37

- 19 -

A P P E N D I X **D**

1992 Federal Income Tax Rates

Corporations   Individuals

Regular Tax

Taxable Income	Rate %	Taxable Income	Rate %
$ 0 - 50,000	15	$ 0 - 35,800	15
50,001 - 75,000	25	35,801 - 86,500	28
75,001 - 100,000	34	Over 86,500	31
100,001 - 335,000	39		

Alternative Minimum Tax

Under $40,000	20%
Over $40,000	24%

A P P E N D I X E

The Limited Liability Company: A Possible Form of Entity for Petroleum Exploration*

Despite the current limitations on the availability of tax losses to U.S. oil companies, tax benefits may, nevertheless, be a significant economic benefit during the life of the project, and all possible steps should be taken to preserve the availability of such items for U.S. income tax purposes. For a number of years, taxpayers have sought a form of business entity allowing limited liability to all its members, while at the same time enabling them to use the entity's tax attributes for U.S. income tax purposes.

Historically, some U.S. oil companies used limitadas for foreign operations. Such entities have corporate characteristics of limited liability to all members, but otherwise resemble partnerships. Limitadas are generally found in the laws of several Central American, South American, and Western European countries, but until recently a similar form of entity has not been available in the United States. The type of limitada frequently utilized by U.S. taxpayers in the past was the Panamanian Limitada. Limitadas were superior to the traditional U.S. limited partnerships for the conduct of high-risk ventures, since, under Panamanian law, none of the members of the organization had any liability for debts of the organization except to the extent of capital contributions. Hence, there was no general partner with unlimited liability, as was the case in limited partnerships. The Internal Revenue Service recognized, at least

---

* Revised from Appendix E of Burke and Dole, Business Aspect of Petroleum Exploration in Non-Traditonal Areas (1991).

under the laws of some countries, that the limitada form of entity would qualify as a partnership for U.S. tax purposes and privately ruled on several occasions that such limitadas, including Panamanian Limitadas, would be classified for U.S. income tax purposes.

Despite the U.S. tax treatment of at least certain limitadas as partnerships, such entities had certain disadvantages which prevented their widespread use by U.S. taxpayers. In addition to restrictions on the amount of capital in some cases, the restriction on membership to only natural persons and the uncertainty concerning the grant of immunity for the organization's debts to its members should a suit be brought against a limitada in a U.S. court were viewed as significant impediments in many cases.

For many years, several states have had a type of entity called a "limited partnership association" or "partnership association." In some cases, those types of entities could qualify as a partnership for Federal income tax purposes. Such entities, however, never attracted significant attention from taxpayers due to certain restrictions which affect their usefulness. The primary disadvantage appears to be the uncertainty created by the requirements in the various states' enabling legislation that either the association's principal base of business or its principal office be maintained in the state of organization. Accordingly, if no, or only minimal, activities were conducted in the state of organization, the limited liability feature of such an organization might not protect its members against the organization's liabilities.

Many taxpayers utilize S Corporations for high-risk ventures and now may find such corporations attractive for operations due to the elimination of the restriction on foreign receipts. However, in petroleum exploration activities, it is often necessary to have special allocations among the partners; and, in some cases, indebtedness is incurred (which under S Corporation rules cannot be added to the basis of the individual owners). If special allocations (which are not allowed in S Corporations) and/or liabilities significant to creating tax deductions will be utilized, the S Corporation may not be an appropriate entity for petroleum exploration.

In order to provide a type of entity which would facilitate the availability of limited liability and the availability of partnership treatment for U.S. income tax purposes, the Wyoming legislature enacted the Wyoming Limited Liability Company in 1977. See Burke, Frank M., Jr. and John Sessions, "The Wyoming Limited Liability Company: An Alternative to Sub S and Limited Partnerships?" *The Journal of Taxation 54*, no. 4 (April 1981): 232-235. Several other states, including Florida and Colorado, have enacted very similar statutes. However, the Wyoming Limited Liability Company will be the focus of this discussion, since that was the first limited liability company similar to a limitada created in the United States.

As the name of the organization implies, a Wyoming Limited Liability Company has the corporate characteristic of limited liability for all of its members. If a Wyoming Limited Liability Company (or a limited liability company formed in another state) is to operate in a state or country other than the state of formation, one should be careful to take all steps necessary to assure that operations in other states or countries will be protected under the limited liability concept. It may be necessary to register a Wyoming Limited Liability Company as a corporation in other states in order to attempt to preserve the limited liability characteristic with respect to operations in such state.

In addition to the corporate characteristic of limited liability, Wyoming Limited Liability Companies may have centralized management, depending upon the entity's Articles of Organization. The Wyoming act allows members of a Wyoming Limited Liability Company to provide for the appointment of managers for the company with either continuing exclusive management authority, or authority to act only as the agent for the members. If no manager is named in the Articles, or if the Articles provide for only an agency relationship between the manager or managers and members of the organization, the company should not be considered to have the corporate characteristic of centralized management. On the other hand, if a manager or managers are appointed with continuing exclusive management authority, then the corporate characteristic of centralization of management will probably exist.

An entity formed under the Wyoming act should not have the corporate characteristic continuity of life. Such characteristic is not present because the Wyoming act provides for the dissolution of the entity upon the death, retirement, resignation, expulsion, bankruptcy, or dissolution of a member. Under the Wyoming act, the members have an option to reserve to themselves the right to reconstitute the organization by unanimous, written vote of the remaining members upon the occurrence of one of the events causing dissolution. If such option is desirable, it should be included in the Articles of Organization. The presence of such option should not result in the corporate characteristic of continuity of life not being present, and the Internal Revenue Service has recognized that fact.

Entities formed under the Wyoming act should also lack the corporate characteristic of free transferability of interests, since, under the Wyoming act, a member can transfer or assign all his interest only as provided in the operating agreement. However, unless such transfer or assignment is approved by the unanimous, written consent of the other members of the entity, the transferee will be entitled to receive his share of the profits or other compensation by way of income, or by way of return of contribution. This restriction on transferability precludes the corporate characteristic of free transferability of interest from being present. Since an entity formed under the Wyoming act should lack at least two of the four corporate characteristics discussed above, it should not be deemed to constitute an association taxable as a corporation, but rather should be treated as a partnership for U.S. income tax purposes.

The Internal Revenue Service issued a private ruling in 1980 confirming that a Wyoming Limited Liability Company would be treated as a partnership for Federal income tax purposes, but immediately suspended issuance of any further rulings on the subject. After a period of uncertainty due to reconsideration of a number of partnership issues during the early 1980s, the Internal Revenue Service in 1988 issued Revenue Ruling 88-76, which confirmed that a Wyoming Limited Liability Company should qualify as a partnership for U.S. income tax purposes.

While the limited liability company, whether formed in Wyoming or some other state, may not be the appropriate entity for a particular petroleum project, it is an entity which offers significant flexibility to U.S. taxpayers and should be at least considered as a possible form of entity in petroleum exploration ventures.

I N D E X

A

B

Depreciable assets/basis,
40–41, 101–103

Depreciation, 28, 101–105:
depreciable equipment,
101–103; unit–of–produc-
tion, 103–104; asset transfer,
104–105; recapture, 105

Detonation, 12

Development. *See* Exploration/
development.

Development drilling/well, 19

Devonian shale gas, 130

Discovery value depletion, 5

Disincentive for reserves
replacement, 6

Dispositions (oil and gas
properties), 22, 145–152:
sublease, 146–147; sale,
147–148; nonoperating
interest sale, 148–149;
production payments,
149–151; abandoned/worth-
less properties, 151–152

Division order, 19, 21

Domestic energy tax policy,
176–183

Domestic enhanced oil recovery credit, 132–133

Drilling platform, 96–97

Drilling rig, 17

Drillsite selection/preparation,
17

Dry hole contribution, 16,
54–56

Dual completion, 18

E

Economic interest, 30, 46, 59,
114, 167–168

Enhanced oil recovery credit,
132–133

EOR certification, 132–133

Equalizations, 74–76

Equipment cost, 28

Excess IDC, 140–141

Expenditures treatment, 2–3

Exploration program, 52

Exploration rights, 12

Exploration/development,
16–20

Exploratory drilling/well, 12,
20

F

Fair market value, 5

Farmee, 16, 60

Farmor, 16, 60

Farmout, 16–17, 26, 59–63

Federal income tax rates, 244

Fee interest, 12–13

Fee royalty, 14–15

Fee simple, 47

Financial risk, 2, 5–8, 26

Fixed contract gas, 119–120

Florida, 248

Flowline, 18–19

Footage rate contract, 17

Foreign income tax/tax credit,
164, 166–167

Injection well, 98–99, 110

Input well. *See* Injection well.

Installation cost, 95–96

Intangible drilling/development costs, 31, 87–105, 168–169: tax preference for IDC, 90–91; integrated companies, 91; payout concept, 91–93; IDC election, 93; IDC deduction timing, 94–95; IDC identification, 95–96; IDC offshore, 96–97; service wells, 98–99; IDC/depletion recapture, 99–100; capitalization rules, 100–101; turnkey drilling contracts, 101; depreciable equipment, 101–103; unit-of-production depreciation, 103–104; asset transfer, 104–105; depreciation recapture, 105

Integrated companies (IDC), 91

Interest transferability, 158

Interest–free loan, 82–83

International operations (U.S. taxpayers), 163–170: foreign tax credit, 164, 166–167; economic interest, 167–168; intangible drilling/development costs, 168–169; percentage depletion, 169; geological/geophysical costs, 169–170

Investment credit, 130

IPRO. *See* Independent producers and Royalty owners exemption.

IRS position (geological/geophysical costs), 52–55

J

Joint operating agreement, 19

Joint profit objective, 157

Jurisdiction, 13

L

Labor cost, 21

Landowner's royalty, 14–15, 24, 27, 30, 39, 58

Lease abandonment, 22

Lease bonus, 14, 20, 39–40, 116–117: depletion, 116–117

Lease conveyance, 22

Lease equipment, 18

Lease inventory, 50

Lease sale, 22

Lease supervision, 21

Lease termination, 14

Leasehold cost, 39, 50, 61, 70

Leasing transaction, 38–39

Legal fees, 47

Lessee rights/obligations, 14, 26

Lessor rights/obligations, 14

Limitada, 246–250

Limitations on deductions, 135–143: nonrecourse debt, 136–137; at–risk rule, 138; passive loss limitations, 138–140; alternative minimum tax, 140–143

Short sale, 83–84

Shut-in royalties, 15, 45

Site preparation, 88

Soil analysis, 50

Spudding, 17, 108

Steam drive injection, 111

Steam generation, 132

Storage tank, 19

Stripper well, 124

Structural/stratigraphic traps, 10

Subleasing transaction, 38–41, 146–147: developed property, 40–41

Sublessor, 146–147

Subsurface natural gas storage facility, 98

Surface casing, 17

Surface damage, 43

Surface rights, 12

Surveying/mapping/testing, 5

Synthetic coal fuel, 130, 132

T

Take-or-pay contracts, 81–82

Tangible equipment, 101–103, 132

Tar sand oil, 130

Tax abuse, 44

Tax basis, 4, 36–38, 40–41, 61, 74, 104–105, 114, 116

Tax consequences, 36, 82–83

Tax credits, 129–133: nonconventional fuel credit, 130–132; domestic enhanced oil recovery credit, 132–133

Tax incentives (oil and gas income), 4–7

Tax on disposition (foreign ownership), 172–174

Tax policy (U.S.), 176–183

Tax preference, 90–91, 140: IDC, 90–91

Tax property concept, 23–34: property law, 26; mineral interests, 26; overriding royalty, 27; net profits, 27–29; production payment, 29; economic interest, 30; definition, 31–34

Tax shelters, 95, 136

Tax system (history), 4–7

Tax treatment (geological/ geophysical costs), 50–52

Taxable income, 7

Term mineral rights conveyance, 24

Tertiary injectants, 132

Tertiary recovery, 21, 109–111

Tight formation gas, 130

Time guidelines for deferred costs, 54

Top leasing, 46–47

Transferred property, 123

Transportation system, 78

Turnkey drilling, 17, 89, 101: IDC, 101